iPhone Cool Projects

DAVE MARK, SERIES EDITOR

GARY BENNETT
WOLFGANG ANTE
MIKE ASH
BENJAMIN JACKSON
NEIL MIX
STEVEN PETERSON
MATTHEW "CANIS" ROSENFELD

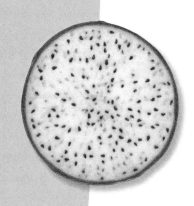

Apress®

iPhone Cool Projects

ISBN-13 (pbk): 978-1-4302-2357-3

ISBN-13 (electronic): 978-1-4302-2358-0

Printed and bound in the United States of America 9 8 7 6 5 4 3 2 1

Trademarked names may appear in this book. Rather than use a trademark symbol with every occurrence of a trademarked name, we use the names only in an editorial fashion and to the benefit of the trademark owner, with no intention of infringement of the trademark.

Lead Editor: Clay Andres
Development Editor: Douglas Pundick
Technical Reviewers: Glenn Cole, Gary Bennett
Editorial Board: Clay Andres, Steve Anglin, Mark Beckner, Ewan Buckingham, Tony Campbell, Gary Cornell, Jonathan Gennick, Michelle Lowman, Matthew Moodie, Jeffrey Pepper, Frank Pohlmann, Ben Renow-Clarke, Dominic Shakeshaft, Matt Wade, Tom Welsh
Copy Editor: Heather Lang
Associate Production Director: Kari Brooks-Copony
Production Editor: Laura Esterman
Compositor: Dina Quan
Proofreader: April Eddy
Indexer: BIM Indexing & Proofreading Services
Artist: April Milne
Cover Designer: Kurt Krames
Manufacturing Director: Tom Debolski

Distributed to the book trade worldwide by Springer-Verlag New York, Inc., 233 Spring Street, 6th Floor, New York, NY 10013. Phone 1-800-SPRINGER, fax 201-348-4505, e-mail orders-ny@springer-sbm.com, or visit http://www.springeronline.com.

For information on translations, please contact Apress directly at 2855 Telegraph Avenue, Suite 600, Berkeley, CA 94705. Phone 510-549-5930, fax 510-549-5939, e-mail info@apress.com, or visit http://www.apress.com.

Apress and friends of ED books may be purchased in bulk for academic, corporate, or promotional use. eBook versions and licenses are also available for most titles. For more information, reference our Special Bulk Sales–eBook Licensing web page at http://www.apress.com/info/bulksales.

The information in this book is distributed on an "as is" basis, without warranty. Although every precaution has been taken in the preparation of this work, neither the author(s) nor Apress shall have any liability to any person or entity with respect to any loss or damage caused or alleged to be caused directly or indirectly by the information contained in this work.

The source code for this book is available to readers at http://www.apress.com. You will need to answer questions pertaining to this book in order to successfully download the code.

Contents at a Glance

Contents

WOLFGANG ANTE

MIKE ASH

GARY BENNETT

MATTHEW "CANIS" ROSENFELD

BENJAMIN JACKSON

NEIL MIX

STEVEN PETERSON

About the Lead Author

Gary Bennett is the lead author on this project. He served for 10 years as a nuclear power engineer on two different nuclear powered submarines. On shore duty, Gary completed his Bachelor of Science degree in computer science.

After college, he worked for GTE Data Services and Arizona Public Service converting hundreds of thousands of lines of OS/2 code to Windows NT. Gary then worked for several technology and health care companies developing Windows NT and Linux applications, including satellite communications. After that, Gary was chief information officer of a young health care company that successfully completed an IPO.

In 2007, Gary started his own technology company, xcelMe.com, focusing on Mac and iPhone development. In 2008, xcelMe.com was hired to develop leading ski and snow report iPhone applications. Since 2008, Gary has been dedicated to teaching others iPhone development. xcelMe.com has developed online iPhone development and marketing courses affordable to all. Gary has taught hundreds of students iPhone development online throughout the world. Gary continues to release helpful iPhone development YouTube videos that benefit the iPhone development community.

In 2009, he worked with EA Sports at their Tiburon studios in Orlando, Florida, where he launched his third iPhone App, Tee Shot Live. He is currently working for a financial institution developing an online banking iPhone app.

About the Technical Consultant

Glenn Cole was the technical consultant on this book. He has been a professional software developer for nearly three decades, from COBOL and IMAGE on the HP 3000 to Java, Perl, shell scripts, and Oracle on the HP 9000. He is a 2003 alumnus of the Cocoa Bootcamp at the Big Nerd Ranch. In his spare time, he enjoys road trips and furthering his technical skills.

Acknowledgments

This book is a compilation of a lot of great work by some really smart authors. They have focused and contributed their chapters based on areas of their expertise. You get to benefit from the years of their expertise; enjoy it!

I am so impressed by the fine people at Apress. I believe their books are the finest on the market. Additionally, they are great to work with. I have made many friends.

I would like to thank "Admiral" Clay Andres whose vision and ability to put together a talented team made this great book possible. He is actually not an Admiral, but should be. Our copy editor, Heather Lang, and development editor, Douglas Pundick, were so very helpful in making sure the quality of the book was what you would want. Special thanks to Laura Esterman and Dina Quan for managing the book production process when it needed it the most.

Lastly I would like to thank Michelle Lowman for connecting Clay Andres and myself and giving me the privilege to be part of this great project.

Gary Bennett

I would like to thank Ivan Neto, Benjamin Maslen, and Rafael Cruz for their hard work on Arcade Hockey, and my parents Lillian Cohn and Larry Jackson for their nonstop love and support.

Benjamin Jackson

Introduction

You are going to love this book! I know I do, and I had to read every word of it and check every line of code, twice!!

If you're like me, you've registered as an iPhone developer with Apple, read some documentation, and sought help in taking the next bold step. Perhaps you've picked up "Beginning iPhone Development," dutifully working through all of the projects, and you understood most of it. If not, I heartily recommend it. The book is great because it gently guides you through many of the technologies that make up an iPhone application. Make no mistake; the book covers a lot of ground. But the projects are kept relatively simple to keep the lessons focused.

First step taken, now boldly onward into the fray!

This book picks up where "Beginning iPhone Development" leaves off. The projects herein were developed specifically for this book, but these are no lightweight applications. Some projects are based on shipping products, showing how various technologies are integrated into a cohesive application. Other projects cover difficult topics and thus are more focused.

The projects illustrate advanced topics such as game timers, XML parsing, streaming audio, multithreading, recognizing advanced gestures, and even designing your own network protocol using UDP (and why you would want to do this). You'll be discussing mutexes, race conditions, sockets, packets, and endianness in no time!

Those who want to develop immersive games have long heard that using a game engine is important, but getting started has been a challenge. Here at last is a game that is built around the open source cocos2d game engine, explained in great detail.

All the chapters represent the personal experience of successful developers; they are written by the developers whose skills we admire and respect.

In short, your next steps are clearly laid out for you.

Who this book is for

This book is for all iPhone and iPod touch developers who want to know more so that they can tackle more difficult programming tasks on their way to creating the next great app. Perhaps you have completed an introductory book such as "Beginning iPhone Development" by Dave Mark and Jeff LaMarche, or possibly you have already completed a simple app and you're ready for the next step on your journey.

It also helps to be comfortable with Cocoa Touch, basic Xcode tools, and Objective-C. You can pick up extra help from "Learn C on the Mac" by Dave Mark, "Expert C Programming" by Peter van der Linden, and "Learn Objective-C for the Mac" by Mark Dalrymple and Scott Knaster.

Mostly, this book is for anyone who wants to write better apps for iPhone and iPod touch and is willing to put in a little time to learn from some of those who have already succeeded at it.

What's in the book

We open with Wolfgang Ante, the developer behind the Frenzic puzzle game, showing how the game was developed and guiding us through the process of creating a similar game called Formic. Timers, animation, and intelligence are used to make the play engaging. If you have been wanting to write a game but have had difficulty getting started, this chapter will provide the guidance and inspiration you need!

Chapter 2 finds Rogue Amoeba's Mike Ash explaining how to design a network protocol using UDP, and demonstrating its use in a peer-to-peer application. This topic is not for the faint of heart, but Mike explains it in a way that makes sense to us mere mortals. I had never seen this topic covered before, so I'm thrilled to see it here.

Next up with Chapter 3 is Gary Bennett covering the daunting but important task of multi-threading. The CPUs in the iPhone and iPod touch won't be mistaken for those of the Mac Pro, but they pack enough power that frequently they are waiting for something to do. Multithreading can be used to keep the user interface responsive while working on other tasks in the background. Gary demonstrates how to do this, and highlights traps to avoid along the way.

In Chapter 4, Canis Lupus (a.k.a. Matthew Rosenfeld) describes the development of the Keynote-controlling application Stage Hand, how the user interface evolved, and the lessons learned from that experience. This knowledge is then demonstrated in a project showing how to recognize many complex gestures at once, including flicking (with inertia!) and rotating an object. Remote controls should all be this handy.

Benjamin Jackson introduces us to two open source libraries in Chapter 5: cocos2d for 2D gaming, and Chipmunk for rigid body physics (think "collisions"). He describes the development of Arcade Hockey, an air hockey game, and explains some of the code used for this. Benjamin then guides us through the creation of a miniature golf game. It's definitely helpful to have such clear guidance through these very murky waters.

Processing streaming audio seems like yet another black art. Luckily for us, Neil Mix of Pandora Radio reveals the science behind the magic in Chapter 6. How do you debug what you can't see? Neil guides us through the toughest challenges, sharing his experience of what works and what to watch for. Audio is hard; I'm thankful to have such a difficult topic explained so clearly. Some of the techniques shown can be used for non-audio applications as well.

The book concludes with Steven Peterson demonstrating a more prosaic integration of iPhone technologies. He weaves Core Location, networking, XML, XPath, and SQLite into a solid and very useful application. Games are great fun, but this is the type of application that makes the device so compelling for the non-gamer. You've seen some of the pieces before; now you'll see how to put them all together.

Software development can be hard. Introductory books lay the foundation, but it can be challenging to understand where to go next. This book shows how to integrate the pieces into a complete application. In addition, many of the topics covered here are notoriously difficult. You'll want to read the chapters more than once, then keep them handy for reference.

Working through the chapters was great fun, and I learned a tremendous amount. I'm sure you will as well!

Glenn Cole

Wolfgang Ante

Company: *ARTIS Software*

Location: *Vienna, Austria*

Former life as a developer: Macintosh Software Developer since 1994. Received the Macworld Editor's Choice Award (1999) and MacUser Award 2004, both for Best Graphics Utility.

Life as an iPhone Developer: Built the Frenzic puzzle game with Xcode and Interface Builder

What's in this chapter: After providing some insight into the development of Frenzic, this chapter discusses a similar game called Formic that shows the basic techniques behind the game logic and animations of a puzzle game.

Key technologies

- *Using* `UIView` *animations for visual feedback*
- *Using* `NSTimers` *to keep the game running*
- *Using* `NSUserDefaults` *to save and restore games*

Designing a Simple, Frenzic-Style Puzzle Game

his chapter is about Frenzic, a popular puzzle game created by ARTIS Software and the Iconfactory. We'll begin by telling you the story behind Frenzic and discussing the design process and some things learned while we developed the game. Finally, we'll guide you through creating a game called Formic, which will demonstrate some of the concepts used in Frenzic.

NOTE

If you do not know Frenzic, head over to http://frenzic.com to download it and see about it for yourself. The version for the iPhone will cost you $2.99, but a version for the Mac that you can download and try for free is also available.

Creating Frenzic

First, let's talk a bit about its history. Frenzic is quite old, I have to confess. I had the basic idea for Frenzic about 18 years ago, while watching a cheesy game show similar to *Wheel of Fortune*. The show involved spinning a big wheel with a ball inside that landed on money values for contestants to win prizes—something clicked, and the basic idea for Leblon (the original name of Frenzic) was born. Initially, there were no power-ups and purely random pies. The game evolved, was ported over to several computer platforms, and got a bit better on every step of the way. You can see its current incarnation in Figure 1-1.

There were two major milestones in advancing the game play: ideal games and power-ups.

In early versions of the game, players felt that, late in the game, they would get unfair pies that they could not set. The pies were chosen randomly, so even if they played a perfect game, players could get pies that made them lose lives. Wolfgang Sykora had the idea to let the application itself play an ideal game in the background, with the same pies you get. An 'ideal game' means clearing circles as soon as possible. Based on this ideal game, players would never be given a pie that could not be set. This made a huge difference! If players try to clear pies as soon as possible and don't make mistakes, they can now possibly play forever if they are fast enough (though, at times, players may decide instead to take risks by filling circles with pie pieces of a single color to potentially win a life).

The second big improvement to game play came when I showed the game to Gedeon Maheux of Iconfactory. He invented the three power-ups that further improved the strategy of the game. Now, players can

Figure 1-1. *The game screen of Frenzic*

take even more risks by filling pies with pieces of a single color in one of the three dedicated power-up circles. Activating the power-ups later allows players to keep the game going even longer and play even faster.

Apart from the game play innovations, several other things have been crucial to the success of Frenzic, the biggest one is my partnership with Iconfactory. Most of the time ARTIS Software is just me, though my wife Arta helps me a lot with testing (she will break, in record time, any code that is not ready). Apart from this, I am a single developer working from my home office, which I love, but being truly successful would require me to be good at all the things that make up great software. Most people, and that includes me, will not be able to do everything well alone. So finding someone who would complement my skills was very important. I have been very lucky to find these partners in Iconfactory. They are some of the best designers in the field of icon and user interface design, and it's an honor to work with them. While I did all the programming on Frenzic, Gedeon Maheux designed the user interface, and David Lanham created the beautiful artwork. The extensive web site was crafted by Anthony Piraino, while Craig Hockenberry and I wrote the code behind it so it works even under heavy load. Last, but not least, Dave Brasgalla did the wonderful music and sound effects.

The web site is a very important part of Frenzic—it may be the most comprehensive high-score list ever created. It also includes player cards that can be customized, player statistics, comments, and different ways to compete: against time (called devotion), against friends, or locally (using the GPS location from the phone). At the time of this writing, more than one million scores are recorded on the Frenzic server. Access to the global high-score tables is possible from the web site as well as from inside the application (see Figure 1-2), so we had to implement web services to communicate with the application and secure the submission of scores to the server to prevent script kiddies from cheating. The whole high-score system amounted to about half of the work that went into Frenzic.

Figure 1-2. *The high-score screen of Frenzic*

Introducing Formic

In this chapter, I want to show you a few of the things Frenzic does. For that, I have created a slimmed-down game called Formic, shown in Figure 1-3. Instead of just showing you some snippets from Frenzic's code, I want to show you a complete game that you can compile, run, and even modify. In the following sections, I will explain the game logic and game graphics in more detail, but I assume some basic knowledge of Xcode and Cocoa.

Like Frenzic, Formic has a middle circle where you will get pieces that you can move to the surrounding circles by tapping on them. The pieces have distinctive shapes. If the center circle's shape matches a surrounding circle's shape, you can move the center piece to the outer circle, and both pieces will be moved out and replaced by new ones. Pieces also have a color, and when you bring together pieces of same shape and color, you win a point. The time to decide where to move a piece is limited and gets shorter the longer you play. If you cannot place a piece in the given time, you lose one of your five lives. The game is over when you have lost all your lives.

Formic is a great project to demonstrate a few things. This very simple and complete game is somewhat similar to Frenzic, but not as much fun. It lacks sound, but it is fully animated and persistent (when a phone call comes in, or you simply quit it by pressing the home button, the game will remember its state and offer to continue the game where you left it on the next launch).

Figure 1-3. *The game screen of Formic*

> **NOTE**
>
> The complete source code of Formic is included on this book's Source Code page of the Apress web site. I have tried to keep it extremely compact and still contain a complete game. There are a few things missing, like sound, but overall, it is a complete game.

Exploring the Formic Code

Formic uses pure and simple Cocoa Touch. It uses NSTimer for scheduling and UIView animations for its graphic effects, just like Frenzic. If you want to write a graphic-intense game, you should probably take a look at OpenGL ES, but for simple puzzle games that just move around a few pieces, this approach is the way to go in my opinion. Nonetheless, keep in mind that Core Animation was built for simple, single animations: it is optimized for ease of use, not for performance. If you decide to use UIView animations or Core Animation, be sure to write some test code that simulates the most demanding animation your game will probably face, and don't forget to also play sound. Don't wait and add sound at the end, as playing music and sound effects on the iPhone does consume noticeable amounts of processing power. Playing sounds has to be part of the simulation.

It also uses the classic Model View Controller (MVC) pattern in a loose way, where the model would be the game object. Figure 1-4 shows a basic MVC pattern.

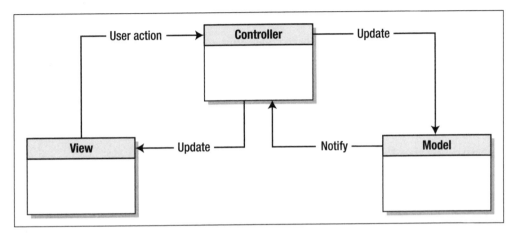

Figure 1-4. *The classic Cocoa MVC flowchart*

The views themselves are quite dumb: they just know how to display themselves. Most of them are simple UIImageViews, with one exception—the background view draws the circles and knows about their positions. Therefore, it also accepts the taps and translates the coordinates back into the tapped circles. This input is then sent directly to the game object, bypassing the controller. The main view controller is responsible for keeping all the views together and animating them. The game logic is isolated in a model object; it keeps the game running and talks to the view controller to make the state of the game visible. This layout leads to the updated flowchart for Formic's objects shown in Figure 1-5.

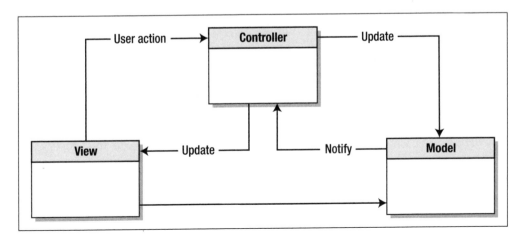

Figure 1-5. *The MVC flowchart for Formic*

You should always try to keep the game logic and graphics separate, though it is sometimes difficult to keep them 100 percent apart from each other. But keeping these functionalities in different objects will make it easier to adapt and fine-tune the game, which is something that will take a lot of the total development time of your game. Good games are not created on the drawing board; you have to play them to see what's great and what's not, and alter accordingly.

In the following sections, you will learn to create Formic. We'll starting from an empty project and create the game object that contains all the game logic, the view controller that keeps all the views together and animates them. Finally, we'll create the custom view that sits in the background of all the views, accepts the player's taps, and converts them into logical taps for the circles that are directly fed into the game object.

Setting Up the Project

Before starting to write code, you need to set up a project. From Xcode's **File** menu, choose **New Project**, and chose **View-Based Application**, as shown in Figure 1-6.

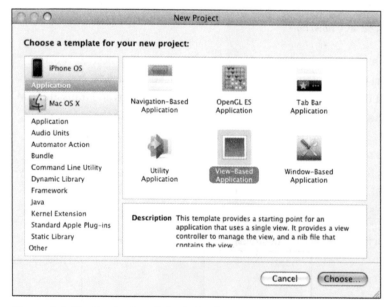

Figure 1-6. *The New Project dialog in Xcode*

This will create a basic project that has a lot of things already set up for you. This simple command created the complete structure of the application, so the only thing left is to create the game object. In this example, I called the project Formic, so it set up the source files for the application delegate and called them *FormicAppDelegate.h* and *FormicAppDelegate.m*. It did the same for the view controller, which it created inside an Interface Builder file that it called *FormicViewController.xib*. It set up the source files for this too, named *FormicViewController.h* and *FormicViewController.m*. Finally, it set up all the necessary connections in Interface Builder so that you already have a convenient `FormicViewController` variable inside your application delegate.

Although these files are created automatically for you, it's a good idea to take a step back and look at what has been created and where to find it.

The FormicApplicationDelegate is the starting point. When the application has started, it will call the `applicationDidFinishLaunching:` method. This is where the code can get things going like creating the game object.

The `FormicViewController` itself lives inside the XIB file. It will be instantiated by the application at startup. You will find a pointer to your view controller in the application delegate, and you will find empty shells for your view controller source files in your project. Just add your controller logic there.

Finally, the view that has been set up for you already lives inside the XIB file. This simple `UIView` will not display anything. To get something displayed, you will have to create a subclass of `UIView`. To do this, select the Classes group in the project tree and choose **New File** from Xcode's **File** menu, as shown in Figure 1-7.

Figure 1-7. *Xcode's New File dialog to create the UIView subclass*

Call them *FormicView.m* to keep to the naming scheme used so far. The files will be created prefilled with all the code necessary to subclass from `UIView` and added to your project. Add your view code in these files.

To finalize the view, you have to change the class of the view inside the XIB file to FormicView. For this, open the file *FormicViewController.xib*, and select the view. Find the inspector panel (or open it from the menu by selecting **Tools ➤ Inspector**), and click the information icon (or press ⌘4) to change the class to FormicView (see Figure 1-8). Save the change, then return to xScope.

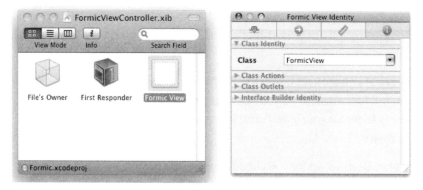

Figure 1-8. *Interface Builder file with Inspector to change the class of the view*

The final step to set up the structure of the application is to create the files for the game object. Click on the Classes group in the project tree, then select the **New File** option from Xcode's **File** menu and create an NSObject subclass called *FormicGame.m*, just as before with the FormicView files. After this, all the necessary objects are created, connected, and ready to be filled with functionality.

Coding the Game Object

Let's start with looking at the game object, because it's the central part of the game. It will talk to the view controller to make the state of the game visible, so it takes a pointer to the view controller in its init method and initializes the game structures. See Listing 1-1.

Listing 1-1. *Initializing the Controller*

```
- (id)initWithViewController:(FormicViewController *)controller
{
    // initialize super
    self = [super init];
    if (!self)
        return nil;

    // general initializations
    mController = [controller retain];
    mLives = 5;
    mTime = 0;
    mPoints = 0;
    mState = GAME_INIT;
```

```
    mBlocked = NO;
    mCenter[GAME_COLOR] = mCenter[GAME_SHAPE] = 0;
    for (int i = 0; i < GAME_CIRCLES; i++)
        mCircle[i][GAME_COLOR] = mCircle[i][GAME_SHAPE] = 0;

    return self;
}
```

The game variables will keep the center and circle shapes and colors, the time left to place the central piece, the points and lives left, as well as the state the game is in. I use the prefix m_ for all class variables. This way they are easily identifiable in the source code (see Listing 1-2).

Listing 1-2. *The Game Variables*

```
int mCenter[2];                 // the color and shape of the center piece
int mCircle[GAME_CIRCLES][2];   // the colors and shapes of the
                                // surrounding circles
int mTime;                      // the state of the running-out timer
int mLives;                     // the number of lives left
int mPoints;                    // the amount of pieces set
BOOL mState;                    // the state of the game (running, over,
                                // etc.)
BOOL mBlocked;                  // if blocked for animations to finish
```

One variable is of special interest here, mBlocked, through which Formic uses the concept of blocking. When animations are going on, the pieces involved will be in an intermediate state. For example, while the piece in the middle is moving out to a circle, the corresponding outer circle piece is still there and will start to fade out as the center piece reaches it. But the game itself does not have intermediate states. When the center piece has the same shape as in the tapped circle, both pieces will be renewed. Therefore, during the animation, the views and the game logic are out of sync. In that time frame, clicking the circle involved will create weird effects.

This is a general problem, not specific to Formic, and it can be addressed in a couple of ways. The first one would be the totally clean one: pieces going into an animation would be removed from the normal view storage and put into a special animation queue. Also, the view controller could not rely on its own view storage and would have to ask the game object about pieces every time it accesses them. This approach, of course, requires a lot of code and an increase in messaging between the controller and the game.

The second way to deal with this problem is to block the game until the animation is finished (see Figure 1-9). This is much simpler and shorter, but if blocking creates undesired gaps in your game play, you obviously cannot use it.

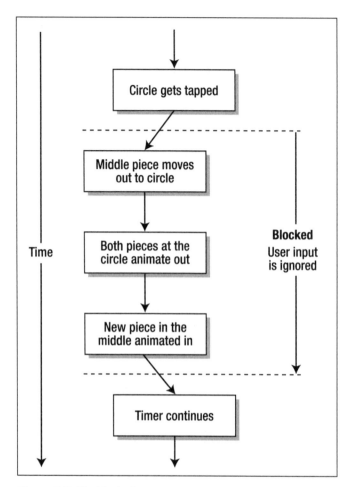

Figure 1-9. *The blocked state*

Formic uses the second, simple blocking approach, and in this case, the blocking is actually a good thing: while the pieces are moving out, its only fair to hold the timer (see the previous discussion about introducing the "timer"), since you do not see your new piece yet.

After the initialization, the game object is in a waiting state. As soon as you tap the center circle, the game will be started by the startGame method. See Listing 1-3.

Listing 1-3. *The startGame Method*

```
- (void)startGame
{
    // don't start over
    if (mState == GAME_RUNNING)
        return;
    mState = GAME_RUNNING;
```

```
    // tell the controller about it
    [mController startGame];

    // fill the outer circles
    for (int i = 0; i < GAME_CIRCLES; i++)
        [self performSelector:@selector(newPieceForCircle:)
withObject:[NSNumber
            numberWithInteger:i] afterDelay:((float)i*0.2)];

    // fill the inner circle
    [self performSelector:@selector(newCenterPiece) withObject:nil afterDe-
lay:1.4];

    // let the game begin
    [self performSelector:@selector(startTimer) withObject:nil afterDe-
lay:1.6];
    [mController updateLives:mLives];
}
```

The startGame method fills the outer circles with shapes and gives you the first piece in the middle. After that, it starts the game timer to get the game going.

The most interesting aspect of this code follows:

```
(void)performSelector:(SEL)aSelector withObject:(id)anArgument
afterDelay:(NSTimeInterval)delay;
```

This method is part of the functionality of NSObject, and it allows you to schedule the execution of a method at a later time. It's extremely easy and flexible to use—just tell the object itself which method to call, when, and with what argument.

The startGame method is used to create the introductory animation, where the pieces around the circle are moved in one after the other, and the center piece comes in at the end (see Figure 1-10). The starting of the timer is delayed to avoid interfering with this introductory animation. It is started with this method:

```
- (void)startTimer
{
    [NSTimer scheduledTimerWithTimeInterval:[self timerInterval]
target:self
        selector:@selector(timerAdvanced:) userInfo:nil repeats:YES];
}
```

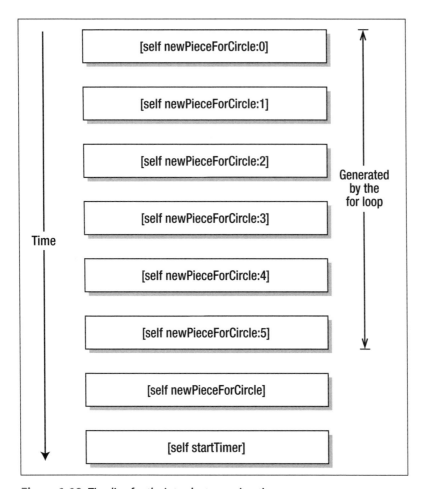

Figure 1-10. *Timeline for the introductory animation*

Note that the delay used for advancing the timer in the game is calculated by the `timerInterval` method. This will create shorter intervals the more points you have scored. While the game goes on, the timer will be restarted after every won point to make the game run faster.

The timer will repeatedly call the method in Listing 1-4.

Listing 1-4. *The Method to Be Called by the Timer*

```
- (void)timerAdvanced:(NSTimer *)timer
{
    // don't advance when blocked
    if (mBlocked)
        return;
```

```
    // new piece, new timing
    if (mTime == 0)
    {
        [timer invalidate];
        [self startTimer];
    }

    // advance timer
    [mController updateTimer:mTime];
    mTime++;
    if (mTime >= GAME_TIMERSTEPS)
    {
        // lost a life
        mLives--;
        [mController updateLives:mLives];
        if (mLives <= 0)
        {
            // game over
            mState = GAME_OVER;
            [timer invalidate];
            [mController gameOver];
        }
        else
        {
            // next piece
            [self newCenterPiece];
            mTime = 0;
        }
    }
}
```

First, note that the previously mentioned game blocking is respected here. If the game is blocked, this method will do nothing.

The next thing is to adapt the timer's interval. The more points the user scores, the faster the game moves. Since timer intervals cannot be changed, you have to delete the old timer and create a new one.

Finally, the time display has to be updated, and the time counter increased. If the player has depleted the allotted time, a life is lost and the center piece is replaced. Once all the lives are gone, the game is over. All this information has to be checked in this method, which is like the heartbeat of the game. Any changes have to be communicated to the view controller.

The other important method of the game object is called when the user taps a circle to move the center piece to it. This method is called by the background view every time a circle is tapped. It returns a BOOL to indicate if the center piece was movable. See Listing 1-5.

Listing 1-5. *The Method Called by Tapping in a Circle*

```
- (BOOL)moveCenterToCircle:(int)circle
{
    // no placement when blocked or game over
    if (mBlocked || (mState == GAME_OVER))
        return NO;

    if (mCenter[GAME_SHAPE] == mCircle[circle][GAME_SHAPE])
    {
        // see if they have the same color
        if (mCenter[GAME_COLOR] == mCircle[circle][GAME_COLOR])
        {
            mPoints++;
            [mController updateScore:mPoints];
        }

        // start moving and create new center
        [mController moveCenterToCircle:circle];
        mCenter[GAME_COLOR] = mCenter[GAME_SHAPE] = 0;
        [self newCenterPiece];
        mBlocked = YES;

        // yes we can!
        return YES;
    }
    else
        // cannot be placed
        return NO;
}
```

Again, this has to respect game blocking and will do nothing if the game blocked or finished.

If the shapes match, an animation is started by a call to the view controller. At this point, the game is blocked. When the animation is finished, it will call the method in Listing 1-6, where we unblock the game again. This way the game is essentially paused for the duration of the animation, and the game and the graphics will stay in sync.

Listing 1-6. *Finding a Suitable New Piece for a Given Outer Circle*

```
- (void)newPieceForCircle:(NSNumber *)circle
{
    int        num = [circle intValue];
    BOOL     centerFound = NO;

    // find new piece, and assure center piece can be set
    for (int i = 0; i < GAME_CIRCLES; i++)
```

```
        if ((mCenter[GAME_SHAPE] == mCircle[i][GAME_SHAPE]) && (i != num))
            centerFound = YES;
    mCircle[num][GAME_COLOR] = rand () % GAME_MAXCOLORS;
    if (centerFound)
        mCircle[num][GAME_SHAPE] = rand () % GAME_MAXSHAPES;
    else
        mCircle[num][GAME_SHAPE] = mCenter[GAME_SHAPE];


    // display it
    [mController zoomInCircle:num withColor:mCircle[num][GAME_COLOR]
            andShape:mCircle[num][GAME_SHAPE]];
    mBlocked = NO;
}
```

This method should simply create a new shape for that circle. The first approach to this problem would be to simply create a random piece. When you do this, you will get to a point where the circles are filled with shapes that the user cannot replace with the center piece and will lose a life.

To keep the frustration at bay, you should always have at least one circle where you could set the center piece. That is what happens in Listing 1-6.

In Frenzic, the method in Listing 1-6 took a very long time to get right. From the beginning, the goal was to make the game as much fun as possible, and the frustrating pies that could not be placed worked against that goal. On the other hand, giving out only pies that could be set would reduce Frenzic to a simple tap-as-fast-as-you-can game with no strategy. In addition to the ideal game that is played in the background, Frenzic uses more rules to give you your pie. The lesson here is: tweaking your game while you develop it is essential.

The method for a providing a new center piece is in Listing 1-7.

Listing 1-7. *Finding a Suitable New Piece for the Center*

```
- (void)newCenterPiece
{
    // fade existing one out
    [mController zoomOutCenter];

    // find a new one
    mCenter[GAME_COLOR] = rand () % GAME_MAXCOLORS;
    mCenter[GAME_SHAPE] = mCircle[rand () % GAME_CIRCLES][GAME_SHAPE];

    // display it
    [mController zoomInCenterwithColor:mCenter[GAME_COLOR]
      andShape:mCenter[GAME_SHAPE]];
```

```
    // reset the timer
    mTime = 0;
    [mController updateTimer:mTime];
}
```

Here, we have to also make sure the user does not get a piece that cannot be set. Most of the time, the method for adding a new piece at one of the outer circles does this already, but if you lose a piece because of a time-out, this method assures the replacement piece is one that can be set.

This completely defines the game logic. The rest is housekeeping functionality, like saving and restoring games (which will be described at the end of this chapter), plus displaying and animating (which are handled inside the view controller).

Coding the View Controller

The view controller manages all the graphics, including the animations. You have seen the calls to the view controller from the game class already, but now, it is time to highlight some of its code.

The following methods are called from the game object directly. Their names should reveal their purposes easily:

```
- (void)zoomInCenterwithColor:(int)color andShape:(int)shape;
- (void)zoomOutCenter;
- (void)moveCenterToCircle:(int)circle;
- (void)zoomInCircle:(int)circle withColor:(int)color andShape:(int)shape;
- (void)updateTimer:(int)timervalue;
- (void)updateLives:(int)lives;
- (void)updateScore:(int)points;
```

All views, except the background view, are simple image or label views. In Frenzic, there is another subclassed image view that has the ability to pulse. Apart from that, just like Formic, Frenzic uses only ready Cocoa views, too.

The methods in the view controller are all very similar, since they use the same basic concept to animate and display views. The principle of these animations is to change a property, like the position, transparency, or size of a view, and to let the change be animated over a given time frame instead of changing the property of the view immediately.

To begin an animation you simply start, with this method:

```
[UIView beginAnimations:nil context:nil];
```

Then, you set the duration of the animation:

```
[UIView setAnimationDuration:DURATION];
```

Next, change the view properties. To start the animation, end your code block with this line:

```
[UIView commitAnimations];
```

Here is an example. To fade in a view, you would set the alpha value of that view to 0.0 and add it to the view hierarchy. Then animate its alpha value to 1.0. This will create a fade in:

```
VIEW.alpha = 0.0;
[MASTERVIEW addSubview:VIEW];
[UIView beginAnimations:nil context:nil];
[UIView setAnimationDuration:ANIM_NORMAL];
VIEW.alpha = 1.0;
[UIView commitAnimations];
```

The following view properties can be animated:

- `frame`, `bounds`, and `center` (i.e., position)

- `alpha` (i.e., transparency)

- `transform` (i.e., size)

All of these animations may be combined, which will create the rich animations you see in Formic (and Frenzic).

Sometimes, Formic includes complex, stacked animations. One of these is moving the piece from the center to the desired outer circle position and then fading out the matching pieces as new pieces fade in. In Frenzic one of these is when the pies move out to the circle, slightly lifted, then set down when over the circle.

For these complex animations, you have to do several animations one after another. One way to do this is to use the scheduled execution of a method, as I already mentioned:

```
[self performSelector:@selector(animationDidStop:)
  withObject:PARAMETER afterDelay:DURATION];
```

Another way to do this would be to set the animation delegate and selector in the animation block between `beginAnimations` and `commitAnimations` like this:

```
[UIView setAnimationDelegate:self];
[UIView setAnimationDidStopSelector:@selector(animationDidStop:finish:)];
```

Since this takes two lines of code and has an additional `finish:` parameter that isn't very useful most of the time, I have always preferred to use `performSelector:WithObject:afterDelay:` instead.

Listing 1-8 contains the two methods from Formic that demonstrate these stacked animations in real code.

Listing 1-8. *Two Methods Forming a Complex Animation*

```
- (void)moveCenterToCircle:(int)circle
{
    // animate it there
    [UIView beginAnimations:nil context:nil];
    [UIView setAnimationDuration:ANIM_NORMAL];
    mCenterView.alpha = 1.0;
    mCenterView.transform = CGAffineTransformMakeScale (0.95, 0.95);
    mCenterView.center = [(FormicView *)[self view]
centerForCircle:circle];
    [UIView commitAnimations];

    // transfer and schedule finishing up
    mMovedView = mCenterView;
    mCenterView = nil;
    [self performSelector:@selector(clearCircle:) withObject:
      [NSNumber numberWithInt:circle]afterDelay:ANIM_NORMAL];
}

- (void)clearCircle:(NSNumber *)number
{
    int    circle = [number intValue];

    // animate inner and outer piece out
    [UIView beginAnimations:nil context:nil];
    [UIView setAnimationDuration:ANIM_NORMAL];
    mMovedView.alpha = 0.0;
    mMovedView.transform = CGAffineTransformMakeScale (0.33, 0.33);
    mCircleView[circle].alpha = 0.0;
    mCircleView[circle].transform = CGAffineTransformMakeScale (3.0, 3.0);
    [UIView commitAnimations];

    // and remove them
    [mMovedView release];
    mMovedView = nil;
    [mCircleView[circle] release];
    mCircleView[circle] = nil;

    // then move new piece in
    [[AppDelegate game] newPieceForCircle:[NSNumber numberWithInt:circle]];
}
```

Be creative with the use of these animations. For example, you could double views just to add some effects on top of the real view. The points and lives in Formic look like they're flying out when you score a point or lose a life. This effect is accomplished by adding a copy of the view on top and animating it to increase in size and fade out at the same time.

Coding the Background View

The background view of Formic is where the playing field is drawn, and because it knows the positions of the circles, this is the view that accepts the taps and sends them to the game. Figure 1-11 shows the Formic background view.

Figure 1-11. *The background view draws the circles and accepts the taps.*

Although this background view is very straightforward, it breaks the clean MVC structure. The view has to know about the game, which acts as the model here. In cases like this, I tend to be practical and just let the view tell the game about the tap. The only problem left is the wiring—when do these objects learn about each other? My solution is to not tell objects about one another at all; I give the application delegate that creates an object a @property for this object that all the other objects can read. This solution is convenient because every object can already find out about the application delegate. To make the code easy to read, I define an AppDelegate like this:

```
#define AppDelegate    (FormicAppDelegate *)
                            [[UIApplication sharedApplication] delegate]
```

In the header file of Formic's application delegate, I add the property:

```
@property (readonly) FormicGame *game;
```

Having added these two lines of code, it's easy to call methods of the game object from any-where in the application without setting up any direct connections between the objects. In the background view's method that handles the taps, it looks like this:

```
[[AppDelegate game] moveCenterToCircle:i];
```

Adding iPhone-Specific Functionality

Your iPhone application has to handle a few additional things beyond game play:

- Activation and deactivation (when waking the device and putting it to sleep)

- Memory warnings (for memory shortages of the device)

- Saving and restoring the game (when quitting and opening the application)

Activating and Deactivating Formic

Activation and deactivation notifications are sent to your application delegate with these two calls:

```
- (void)applicationWillResignActive:(UIApplication *)application;
- (void)applicationDidBecomeActive:(UIApplication *)application;
```

These methods are called when the user presses the pause button on top of the device. When you receive applicationWillResignActive: you should stop all timers, animations, and sound, and your game should go into a pause state. The device will not really sleep; it will just go into a power-saving mode and turn the screen off. Music will continue to play, and animations will even continue to run but will not be visible. This power-saving mode will still drain the battery, so you have to handle these. Version 1.0 of Frenzic could drain the battery of your phone in record time when you put your device into sleep, since we forgot to stop some animations that nobody every saw, except the battery.

Adding Memory Warnings

Memory warnings will occur when the device runs out of memory. They are sent to all view controllers in the application in the form of this method:

```
- (void)didReceiveMemoryWarning;
```

You should free as much memory as possible. Think about memory management early because running out of memory is unpredictable and difficult to test. While you almost never run short of memory on your debugging devices, your users will do so under all kind of edge situations. From the crash logs that users sent back to us, we found that most of the crashes of Frenzic's 1.0 version were not actual crashes, but shutdowns. When your application receives memory warnings but fails to free enough memory, the operating system will simply shut down your application, which looks like an application crash to the user.

Also don't think there is a safe lower limit for memory usage. When the device has a memory shortage, your application may not have caused the problem, but your application will be shut down to solve it.

Saving and Restoring the Game

An iPhone application has to be persistent. This means that, although the application can be quit at any time—by incoming phone calls, SMS, or the user pressing the home button—the next time you start the application, it should take off where it was left the last time.

For a game, you should probably offer the choice to continue or start with a new game. In Frenzic, when the game is paused and restored, and the player is given the choice to resume play or start over, as shown in Figure 1-12.

Figure 1-12. *The user is asked if a previously saved game should be resumed.*

Formic uses a simple `UIAlert` but the principle is the same. If the game is quit while not running (i.e., in the initialization or game-over state), there is nothing to save, and therefore nothing to restore.

<u>NOTE</u>

The complete source code is included with this book's source code on the Apress web site. You can run Formic inside the iPhone Simulator and set breakpoints to see in detail how objects interact and when methods get called.

The code for saving and restoring is inside the game class and gets invoked from the application controller. On startup, the applications delegate's method applicationDidFinishLaunching: will be called, and on shutdown, the method applicationWillTerminate: will be called. These are the two points to get the game class to save and restore the game.

The simplest way to store settings is in the standard user defaults. The class NSUserDefaults offers a very simple way to store data persistently; it works like a dictionary. See Listing 1-9.

Listing 1-9. *Saving and Restoring the Game*

```
-  (void)saveGame
{
    NSUserDefaults    *prefs = nil;

    prefs = [NSUserDefaults standardUserDefaults];
    if (mState == GAME_RUNNING)
    {
        // save the data representing the game to the preferences
        [prefs setObject:[NSNumber numberWithBool:YES] forKey:@"saved"];
        [prefs setObject:[NSData dataWithBytes:mCircle
length:sizeof(mCircle)]
forKey:@"circle"];
        [prefs setObject:[NSNumber numberWithInt:mLives] forKey:@"lives"];
        [prefs setObject:[NSNumber numberWithInt:mPoints]
forKey:@"points"];
    }
    else
        // save the 'no game data' indication to the preferences
        [prefs setObject:[NSNumber numberWithBool:NO] forKey:@"saved"];
}

-  (void)restoreGame
{
    NSUserDefaults    *prefs = nil;

    prefs = [NSUserDefaults standardUserDefaults];

    // get the data from the preferences
    [[prefs dataForKey:@"circle"] getBytes:mCircle length:sizeof(mCircle)];
```

```
    mTime = 0;
    mLives = [prefs integerForKey:@"lives"];
    mPoints = [prefs integerForKey:@"points"];
    mState = GAME_RUNNING;

    // fill the outer circles
    for (int i = 0; i < GAME_CIRCLES; i++)
        [self performSelector:@selector(zoomInCircle:) withObject:
            [NSNumber numberWithInteger:i] afterDelay:((float)i*0.2)];

    // new inner circle
    [self performSelector:@selector(newCenterPiece) withObject:nil afterDe-
lay:1.4];

    // let the game begin
    [self performSelector:@selector(startTimer) withObject:nil afterDe-
lay:1.6];
    [mController updateLives:mLives];
    [mController updateScore:mPoints];
}
```

Summary

Using standard Cocoa Touch to create a game like Formic for the iPhone is quite unusual. Usually, for graphics-intense applications, you should look into other ways to do your coding, like OpenGL ES. If you are writing a puzzle game with a few effects and animations running at once, using Cocoa Touch is perfectly fine and will help you get things done quickly and with less effort.

Keep in mind, though, that Cocoa was not built for games. Before you start using Core Animation with UIViews and NSTimers, make sure that your final game will not suffer from that decision. Write a prototype and simulate cases that you think will put the most stress on your game. Don't forget to include sound in your tests; sound effects might be just the last piece that will make your game stutter.

Separate game logic and graphics from each other. The game classes in Frenzic for the Mac and iPhone are basically the same, but the graphics and visuals—the whole user interface—is totally different. This will also help you when you start to tweak the game to make it more fun, since all the code you need to change will be in one place.

And finally, pay attention to the iPhone-specific needs of your application. Be especially careful about memory warnings. On my device, I have never seen a single one, but as soon as Frenzic got in the hands of beta testers, warnings started to show up. When you ignore them, the device will shut down your application, and this will look like a crash to your users.

Mike Ash

Company: *Rogue Amoeba Software, LLC*

Location: *Alexandria, VA, USA*

Former life as a developer: **Started out in BASIC on a Commodore 64, and graduated to AppleSoft BASIC on an Apple IIGS and then Pascal. Moved up to a Mac, and did Think Pascal there, then C and C++ with CodeWarrior. Got started with Objective-C in 2000 on Linux, partly to get ready for Mac OS X and Cocoa. These days, mainly does Objective-C in Xcode, writing Mac applications, with some Python on the side for scripting and server-side code. Current specialties are audio, multithreading, networking, and performance optimization.**

Life as an iPhone developer: **The NetAwake application, in the utility category, is currently in the App Store. Also near shipping Nanogolf, a game.**

What's in this chapter: **This chapter covers networking with UDP.**

Key technologies

- **POSIX sockets**

- **UDP networking**

- **Bonjour**

Mike Ash's Deep Dive Into Peer-to-Peer Networking

The iPhone is a remarkable device; it squeezes the power of a serious desktop machine from only a decade ago into a machine that's smaller than a deck of cards. The great enthusiasm for it by users and developers alike shows just how useful such a thing can be. And despite the powerful hardware, the functional design, and the carefully crafted software, it would be nearly pointless to own the only iPhone in the world. Networking is what truly makes the iPhone compelling. Not just having this palm-sized computer but being able to use it to interact with other palmtop computers across the room and across the globe.

Networking has always been an acronym soup. Today's soup is made up of buzzwords like REST, XML, HTTP, JSON, and AJAX. These are great technologies, and they allow wonderful new applications to be built, applications that we all use on a daily basis. However, sometimes it's useful to go back to the old days of pushing raw bytes through a pipe, whether it's for better performance, simplicity, or simply for the challenge of doing something a little different.

Long before these modern technologies existed, the acronym soup was made up of technologies like IP, TCP, and UDP and words like "Ethernet" and "sockets." These provided the low-level infrastructure that all of the modern stuff is built on. That fancy RESTful API that your application talks to is, way down at the bottom of things, using TCP and Ethernet to get the job done. It's always useful to get under the hood and see how things really tick, and I'm going to do a little bit of that here by exploring networking using raw sockets and UDP.

In this chapter, I'm going to walk you through the design of a very simple LAN game. The game itself is secondary, and the primary focus is going to be the networking itself. We'll figure out exactly what we want out of the game, design a network protocol to suit the requirements, and then implement it using straight POSIX sockets.

By the time we've finished, you'll be familiar with the principles of low-level networking and basic binary data encoding and decoding. Whether you're planning to implement a LAN game along similar lines or just want to know how everything works under the hood, it's my hope that you'll enjoy this examination of a more traditional way of doing things.

Planning a Simple Collaborative Game

The game that we're going to build in this chapter is called SphereNet. It's a simple collaborative application that shows a collection of spheres on the screen. When the game is started in isolation, it creates a single, randomly colored sphere that can be moved around the screen by touching or dragging your finger.

When multiple copies are started on the same LAN, each copy displays the spheres from all other copies. The idea is to make the user experience as simple as possible. There's no loading screen, hosting, or joining. The user simply starts the application and uses it, and it automatically searches out and communicates with any other copies of the application that it can reach. The completed game is shown in Figure 2-1.

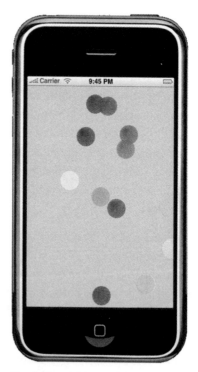

Figure 2-1. *The completed SphereNet game in action*

Building the GUI

I really want to talk about networking, but before we can do that, we need something to actually network. We could just build the protocol directly, but that would be boring. Let's start out by building the graphical user interface (GUI).

We start by creating a new view-based application project in Xcode. The GUI for SphereNet is going to be as simple as possible. It literally has nothing other than a single view. There are no settings and no About screen. The view-based application template fits our needs perfectly.

Let's think about what SphereNetSphere needs to contain. It has to have a color and a position. To keep things fast and smooth, we'll use Core Animation for the display, so we'll give SphereNetSphere a CALayer as well. Last, we're eventually going to need to remove unresponsive networked spheres from the screen, so we need to know when a given sphere was last modified. That covers everything.

Modify *SphereNetSphere.h* so that it appears as follows:

```
#import <Foundation/Foundation.h>


@interface SphereNetSphere : NSObject
{
    // store the color using its red, green, and blue components
    float _r, _g, _b;
    CALayer *_layer;
    NSTimeInterval _lastUpdate;
}

- (void)setColorR:(float)r g:(float)g b:(float)b;
- (float)r;
- (float)g;
- (float)b;
- (void)setPosition:(CGPoint)p;
- (CGPoint)position;
- (CALayer *)layer;
- (NSTimeInterval)lastUpdate;

@end
```

That's nice and straightforward. The initializer is likewise straightforward, with just a few lines to set up the CALayer.

```
static const CGFloat kSphereSize = 40;

- (id)init
{
    if((self = [super init]))
    {
        _layer = [[CALayer alloc] init];
        [_layer setDelegate:self];
        [_layer setBounds:CGRectMake(0, 0, kSphereSize, kSphereSize)];
        [_layer setNeedsDisplay];
    }
```

```
    return self;
}

- (void)dealloc
{
    [_layer release];

    [super dealloc];
}
```

Next come a bunch of boring setters and getters. You'll notice that I'm not using the @property syntax anywhere. This is because I'm an old-timer curmudgeon, and I don't like it very much. Don't worry; there aren't very many accessors in this program.

```
// set the color and update the screen if necessary
- (void)setColorR:(float)r g:(float)g b:(float)b
{
    if(r != _r || g != _g || b != _b)
    {
        _r = r;
        _g = g;
        _b = b;

        [_layer setNeedsDisplay];
    }
}

- (float)r
{
    return _r;
}

- (float)g
{
    return _g;
}

- (float)b
{
    return _b;
}

// set the sphere's position on screen, and note when it occurred
- (void)setPosition:(CGPoint)p
{
    [_layer setPosition:p];
    _lastUpdate = [NSDate timeIntervalSinceReferenceDate];
}
```

```
- (CGPoint)position
{
    return [_layer position];
}

- (CALayer *)layer
{
    return _layer;
}

- (NSTimeInterval)lastUpdate
{
    return _lastUpdate;
}
```

Finally, we need to draw the contents of the CALayer. We'll use CGGradient to draw a radial gradient and give a bit of a 3D effect. The details behind the following drawing are beyond the scope of this chapter. To learn more about the technique used here, see the chapter on gradients in Apple's *Quartz 2D Drawing Guide*, available from http://developer.apple.com.

```
static const CGFloat kSphereCenterOffset = 10;

- (void)drawLayer:(CALayer *)layer inContext:(CGContextRef)ctx
{
    CGFloat locations[2] = { 0.0, 1.0 };
    CGFloat components[8] = { _r, _g, _b, 1.0,
                              _r, _g, _b, 0.7 };

    CGColorSpaceRef colorspace = CGColorSpaceCreateDeviceRGB();
    CGGradientRef gradient = CGGradientCreateWithColorComponents(
                                      colorspace,
                                      components,
                                      locations,
                                      2);

    CGPoint offsetCenter = CGPointMake(
                              kSphereSize / 2 - kSphereCenterOffset,
                              kSphereSize / 2 - kSphereCenterOffset);
    CGPoint center = CGPointMake(kSphereSize / 2, kSphereSize / 2);
    CGContextDrawRadialGradient(          ctx,
                                          gradient,
                                          offsetCenter,
                                          0,
                                          center,
                                          kSphereSize / 2,
                                          0);
```

```
        CFRelease(gradient);
        CFRelease(colorspace);
}
```

Now, let's move on to the view controller. This one is also pretty simple, and the Xcode template has put in the basics for us already. We'll have a single instance variable to hold on to the local sphere.

```
SphereNetSphere *_localSphere;
```

We'll create it when the view loads.

Also, since we're introducing a new class, add @class SphereNetSphere; above the @interface line.

```
- (CGFloat)randomFloat
{
    // generate a random number between 0 and 1
    // random() tops out at 2^31-1, so divide by that
    return (CGFloat)random() / ((1 << 31) - 1);
}

- (void)viewDidLoad
{
    [super viewDidLoad];

    if(!_localSphere)
    {
        _localSphere = [[SphereNetSphere alloc] init];
        CGSize size = [[self view] bounds].size;
        srandomdev();
        [_localSphere setColorR:[self randomFloat]
                              g:[self randomFloat]
                              b:[self randomFloat]];
        [_localSphere setPosition:
                CGPointMake(size.width / 2, size.height / 2)];
        [[[self view] layer] addSublayer:[_localSphere layer]];
    }
}
```

Since we're referencing the layer directly, add #import <QuartzCore/CoreAnimation.h> above the existing #import. Since we're now using SphereNetSphere, add #import "SphereNetSphere.h" as well.

I'm being a little paranoid here by checking to see if the _localSphere variable already exists. The view should load only once, but I'm the paranoid type. It's much easier to add this sort of basic guard than to track down a bug caused by not having it.

That's enough to make the game show up on the screen, but we still need to move the spheres around. Thanks to the magic of CoreAnimation, this is really straightforward too:

```
// if the user touches or drags across the screen,
// update the position accordingly in the animation layer
- (void)moveLocalSphereFromTouch:(UITouch *)touch
{
    if(touch)
        [_localSphere setPosition:[touch locationInView:[self view]]];
}

- (void)touchesBegan:(NSSet *)touches withEvent:(UIEvent *)event
{
    [self moveLocalSphereFromTouch:[touches anyObject]];
}

- (void)touchesMoved:(NSSet *)touches withEvent:(UIEvent *)event
{
    [self moveLocalSphereFromTouch:[touches anyObject]];
}
```

Figure 2-2 shows the completed GUI—that's the basic application, minus networking. It's a small, straight-forward base on which we can build some interesting network code.

Networking the Game

Now that we have a base to start with, we'll move on to building the networking side of SphereNet. First, we'll talk about the goals that we have for the networking code. Next, we'll take those goals and use them to design a custom SphereNet protocol. Finally, we'll take that protocol design and build the actual code that we need to add networking to the game.

Defining the Networking Goals

Before we actually design the network protocol, it's good to have an idea of what the goals for the game are. Knowing what you want out of code before you build it is critical. Without knowing the goals, it's too easy to get lost in complexities or end up with conflicting requirements.

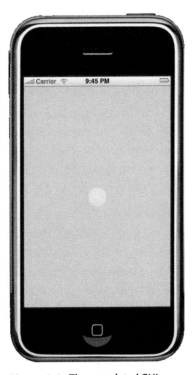

Figure 2-2. *The completed GUI*

The first goal for SphereNet's networking is that it should be simple. Simplicity in any kind of code is always important, both to reduce the amount of work needed to write it and to reduce the potential bugs. It's especially important in networking code because such code necessarily involves two or more machines interacting with each other. Figuring out whether the source of an error is on one machine, another machine, or somewhere in between can be quite difficult and makes networking code about ten times harder to debug than normal code.

The next goal is for the game to be extensible. It should be designed so that we can add new features and behaviors, as much as possible, without breaking older versions of the program. This conflicts with "simple" to some degree but not too much. We will strive to make it be extensible in a simple fashion.

Next, we want the application to be fast. When a user touches the screen, we want other users to see that sphere move instantly.

Last, the application should be platform agnostic. This means using a protocol that not only works on an iPhone but could, at least in theory, work on anything. We'll also need to specify exactly how everything gets encoded in the network packets rather than simply using whatever format is the default for the platform. While this may sound rather theoretical, it's actually quite important even if the code never leaves the iPhone: the iPhone Simulator is different enough from a real iPhone that this is a necessity simply for testing in the simulator.

Designing the Network Code

The protocol itself will be a custom protocol running over User Datagram Protocol (UDP). There are several reasons I decided to use a custom protocol. First, the task at hand seemed simple enough that it would be easier to simply build a custom protocol than try to squeeze into an existing one. Second, a custom protocol keeps overhead to a minimum and performance at a maximum. Finally, it's simply a better educational exercise!

If you're not familiar with UDP, it's one of the two common application-level Internet protocols; the other is Transmission Control Protocol (TCP). TCP is a stream protocol, and it's what's used every time you view a web page, check your mail, or download a file. In essence, TCP creates a bidirectional pipe between two computers and does its best to cover up the unreliable and uncertain nature of the lower-level network that it runs on.

UDP, in contrast, exposes a lot of that uncertainty to the application. It uses a checksum to ensure that no corrupt data is passed through, but otherwise, it makes no attempt to cover up problems. If a router decides to drop a packet, that data simply is never received. If an old packet gets delayed and arrives late, that packet is received out of order. It's up to the individual application to compensate for these problems.

Given this uncertainty, why use it? Simply because UDP uses fewer resources and provides better performance. TCP's connection-based nature means that the connection must be set up and maintained with each remote device that the application is talking to, and this gets expensive if the plan is to support a large number of them, as is the case here. It also can be slower, such as when packets get lost. TCP will recover, but recovery takes time, whereas UDP will just skip over the loss and keep on sending the subsequent updates. If performance is your goal, and if you can deal with losing data, UDP is the way to go. This is why it's used for voice-over-IP applications, online games, and this example project.

We've settled on using UDP, but what exactly is going to be sent over the network? We need to come up with a packet format. I mentioned earlier that we want the protocol to be extensible, and to accomplish that, we'll put a 4-byte type identifier in the header. Unknown type identifiers can simply be ignored, meaning that new versions of the program can send more data using a new type identifier and old versions will just ignore the data they can't understand. For a bit of safety and paranoia, we'll also have a unique 4-byte identifier at the start of the packet that identifies it as being one of ours, and not some wayward packet coming from a completely different program.

Here, then, is the packet header:

```
typedef struct
{
    uint32_t identifier;
    uint32_t datatype;
} PacketHeader;
```

Figure 2-3 shows how this structure will look on the network.

0	1	2	3
0 1 2 3 4 5 6 7 8 9 0 1 2 3 4 5 6 7 8 9 0 1 2 3 4 5 6 7 8 9 0 1
identifier
datatype

Figure 2-3. *The packet header structure as it will appear on the network*

The identifier field will always be the same for any SphereNet packet, and let's define what it will contain:

```
static const uint32_t kSphereNetPacketIdentifier = 'SpHn';
```

It's just a simple ASCII abbreviation of the application's name.

If you haven't seen this sort of thing before, this is just a multicharacter integer constant. It's conceptually similar to a standard character constant such as 'S', except that it builds a longer integer by stringing the characters together. The constant 'SpHn' is simply a convenient way to write 0x5370486E or 1399867502.

The datatype field will then identify the nature of the data that follows this header. Since SphereNet currently sends only position updates, we're only going to have one datatype, but this field leaves open the option to have more in the future. A position update will contain the coordinates of the sphere in question as well as its color.

Now, let's consider how to represent these bits of data. In the application, we represent coordinates using CGPoint and colors with float. Floating-point numbers are inconvenient to deal with in networking, though, because the binary format of a floating-point number is inconvenient. We'll transform everything to integers, which are easier to work with and, when it comes to debugging, easier to read when you're poring over a hex dump of a packet.

The coordinates will just be 32-bit signed integers. This is a bit of overkill, as the iPhone screen is only 320×480, but it adds some future-proofing. When it comes to the colors, there's no point in using anything bigger than a single byte for each color component. That gives us a range of 0–255 for each component, which is already the maximum color resolution that most screens can reproduce. Our position update packet will then look like this:

```
typedef struct
{
    PacketHeader header;
    int32_t x;
    int32_t y;
    uint8_t r;
    uint8_t g;
    uint8_t b;

} PositionPacket;
```

First comes the header which we defined previously, followed by the coordinates, and then the three color components. We'll also need a datatype constant to identify this particular kind of packet:

```
static const uint32_t kSphereNetPositionPacketType = 'posn';
```

One more refinement: when we put these structs into our code, we have to wrap them in #pragma pack(1) and #pragma options align=reset. This is because a C compiler will usually waste some space to gain speed. Computers work fastest when the data they work with is aligned, which is to say that it sits on a memory address that's a multiple of its size. An int32_t is 4 bytes, so the compiler tries to put it on an address that's a multiple of four.

It will also pad out the end of `structs` so that the next `struct` will start on a nice address if it's being used in an array. All of this padding is compiler dependent and has no place being part of a network protocol. The `#pragmas` tell the compiler to stop padding and cram everything into as little space as possible, which is exactly what we want if we're going to be sending stuff out over the network.

Figure 2-4 shows the original packet structure without the `#pragma`, and Figure 2-5 shows the corrected version.

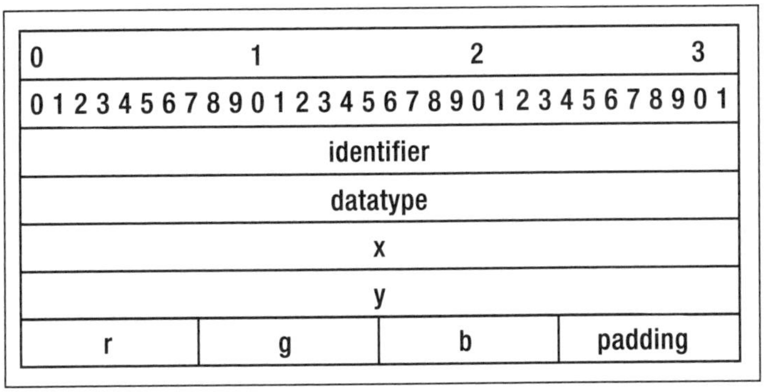

0	1	2	3
0 1 2 3 4 5 6 7 8 9 0 1 2 3 4 5 6 7 8 9 0 1 2 3 4 5 6 7 8 9 0 1			
identifier			
datatype			
x			
y			
r	g	b	padding

Figure 2-4. The packet structure on the network with a typical compiler

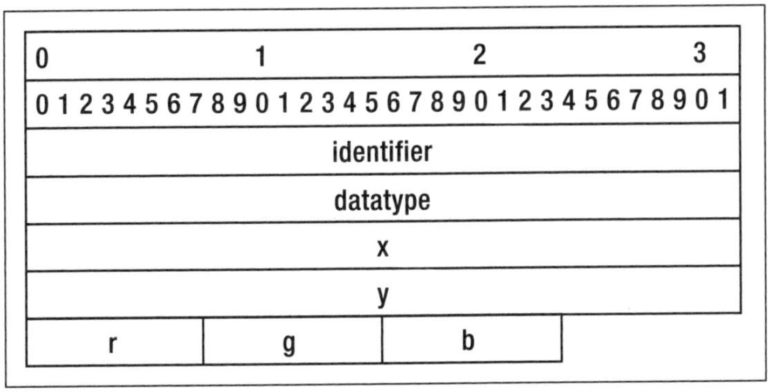

0	1	2	3
0 1 2 3 4 5 6 7 8 9 0 1 2 3 4 5 6 7 8 9 0 1 2 3 4 5 6 7 8 9 0 1			
identifier			
datatype			
x			
y			
r	g	b	

Figure 2-5. The layout of the packet structure with #pragma pack(1) enabled

We don't plan for spheres to change color, so sending the color in every single packet is redundant. Surely, it would be better to simply send the color once and then send only position changes after that?

It's a good idea, but the trouble is the unreliable nature of UDP. If that initial color transmission happened to be lost, the color information would be gone forever. On a typical Wi-Fi network you can count on a roughly 1 percent chance of losing any given packet, so this is a very real possibility! If we did send the color data only with the first transmission, we'd also have to have some way for the other side to acknowledge receipt of the color packet and for the transmitter to resend it if it got lost. Suddenly, the code becomes a lot more complex. It's much simpler, if slightly more wasteful, to just tack 3 bytes onto every position update to ensure that the color is always known to the other side.

Now, you're probably wondering what happens if a position update is lost. After all, doesn't that have the same problem?

The beauty of the position updates is that SphereNet doesn't need to see all of them. If it misses one in the middle, the next one will correct the problem. As a result, the sphere may follow a slightly different path than it did on the sending side, but that's not a big problem. By sending a position update every time the sphere moves, and periodically sending them even when the sphere is not moving at all, we can easily ensure that every other copy of SphereNet stays reasonably up to date with the position of all spheres even in the face of the occasional lost packet.

There's one more piece to the puzzle: figuring out where to send these packets. Apple's Bonjour technology comes to the rescue here. SphereNet can broadcast its presence to the network, and all other copies of SphereNet can find it using Bonjour.

Understanding Endianness

There is one detail that I glossed over in the previous description of the packet format: **endianness**. If you aren't familiar with it, you may be surprised to learn that computers have two different ways to represent integers in memory: big-endian and little-endian. And these two ways are incompatible.

The difference comes about due to the order in which the bytes of a multibyte integer are written. Take the integer 305,419,896 for example. In hexadecimal, this integer would be written out as 0x12345678. The question is, how does it look in memory, for example, as an array of unsigned chars? One obvious way would be to simply write it down in order:

```
unsigned char myInt[4] = { 0x12, 0x34, 0x56, 0x78 };
```

But it's just as reasonable, albeit somewhat less natural, to write it down in the opposite order, with the least-significant byte first:

```
unsigned char myInt[4] = { 0x78, 0x56, 0x34, 0x12 };
```

The former system is called **big-endian**, and the latter is called **little-endian**. The Intel x86 CPU used in Macs these days is little-endian, as is the ARM CPU used in the iPhone. The PowerPC processors used in older Macs are big-endian, and in general, it's common to find either version being used on various platforms. Reading data using the wrong endian will give you garbled, useless numbers, so it's important to get this right.

For historical reasons, big-endian is the de facto standard for transmitting integers over a network and is often referred to as **network byte order** for this reason. As such, we will go with the flow and send all of our integers as big-endian.

NOTE

> There is actually at least one more endianness in the world: middle endian! On some old, rare architectures, neither the forward nor the backward ordering was used, but rather a strange mixed-up ordering that would write the example integer of 305,419,896 as { 0x34, 0x12, 0x78, 0x56 }. The problem of differing endianness is sometimes referred to as the "NUXI problem," due to what happens when storing the string "UNIX" on some of these old systems.

Coding the Networking

To provide clean separation, we're going to put all of the networking code into a separate class, which in a burst of creativity I'm going to call SphereNetNetworkController.

This controller will do two things. First, it will broadcast updates to other controllers when the local sphere is moved, and second, it will notify the main controller of updates received from remote machines. With only two tasks to perform, the public interface is going to be nice and short:

```
- (id)initWithDelegate:(id <SphereNetNetworkControllerDelegate>)delegate;
- (void)localSphereDidMove:(SphereNetSphere *)sphere;
```

We also need to define the delegate protocol. Since there's only one thing it does, notify the delegate of updates, the protocol only has one method.

```
@protocol SphereNetNetworkControllerDelegate

typedef struct
{
    float r, g, b;
    CGPoint position;
} SphereNetSphereUpdate;

- (void)networkController:(SphereNetNetworkController *)controller
        didReceiveUpdate:(SphereNetSphereUpdate)update
        fromAddress:(NSData *)address;

@end
```

Also, since it references SphereNetNetworkController before its @interface line, add @class SphereNetNetworkController; after the #import at the top of the file.

I decided to encapsulate all of the information needed for an update into a struct rather than have a separate parameter for each element, just to keep things simpler and cleaner.

Of course, we're also going to need some instance variables, and to figure out what instance variables are needed, we have to have some idea of how it's going to work.

First off, we need an instance variable to hold the delegate—easy enough.

We also need one to store the socket that's going to be used both to send and receive updates. A POSIX socket is just an int, simple.

Next, we're going to advertise the service using Bonjour, so that means a pointer to an NSNetService that we'll use to do the advertising.

We're also going to find the other SphereNet programs on the local network using Bonjour. This will need a pointer to an NSNetServiceBrowser. We also need to keep track of all the NSNetServices that it returns, which we'll keep in an NSMutableSet.

Last, we need to keep track of the last update that's sent so that we can resend it periodically when there's no activity. The reason for this is that there is no good way to be notified when a remote sphere has left the network. The program could broadcast a "quit" packet as it leaves, but there's no guarantee it would be delivered, just as there's no guarantee any of our packets will be delivered. The answer is to simply time out remote spheres. If no updates are received after some time, remove them from the screen.

To make sure that they don't get removed just because the users finger stopped moving around, we'll resend updates periodically when the program is idle. To do this, we'll store the most recently sent update by borrowing the SphereNetSphereUpdate struct as an instance variable so that it can be resent from time to time.

Here, then, is the complete SphereNetNetworkController interface:

```
@interface SphereNetNetworkController : NSObject
{
    id <SphereNetNetworkControllerDelegate> _delegate;

    int _socket;

    NSNetService *_advertisingService;
    NSNetServiceBrowser *_browser;
    NSMutableSet *_services;

    SphereNetSphereUpdate _lastSentUpdate;
}
```

```
- (id)initWithDelegate:(id <SphereNetNetworkControllerDelegate>)delegate;
- (void)localSphereDidMove:(SphereNetSphere *)sphere;
```

@end

Also, since we're referencing SphereNetSphere, add the forward declaration @class SphereNetSphere; to the top of the file, after the #import line.

Let's get to the code now! We'll start with the initializer, which has to perform a bunch of different tasks. It needs to do basic object initialization, like assigning the delegate ivar, creating the services object, and so on. It needs to create the socket and bind it to a port so that it can receive packets. It needs to advertise the Bonjour service and start a search for other services. And it needs to start up a thread to listen for incoming packets.

The beginning of the initializer is straightforward.

In *SphereNetNetworkController.m*, add the following:

```
#include <netinet/in.h> //for sockaddr_in
#include <sys/socket.h> //for socket(), AF_INET
```

Before we get neck deep in sockets in the following listing, add the necessary #includes above the @implementation line. Also, since we'll be referencing the sphere, import its header as well using #import "SphereNetSphere.h".

```
- (id)initWithDelegate:(id <SphereNetNetworkControllerDelegate>)delegate
{
    if((self = [self init]))
    {
        // assign the delegate
        _delegate = delegate;

        // set up the services container
        _services = [[NSMutableSet alloc] init];

        // set up the UDP socket
        _socket = socket(AF_INET, SOCK_DGRAM, 0);
```

That last call is the POSIX call to make a new socket for network communication. The AF_INET parameter tells it that this is supposed to be an IPv4 socket (as opposed to IPv6, UNIX domain sockets, or other possibilities). The SOCK_DGRAM parameter tells it that we want to use UDP (also known as datagram), and the last parameter specifies a protocol that is not used for UDP, so we pass 0.

Next, we need to bind the socket to a local port. This gets a little strange, as the sockets' API is old and a little weird. The bind() call will bind the socket to an address. An address, in socket terms, is an IP address plus a port. The IP address lets us bind to a particular interface instead of just listening to everything, but there's a special constant to bind to every interface that we'll be using. The address is specified as a `struct`, and there's a different `struct` for each kind of protocol (IPv4, IPv6, etc.). An IPv4 address is specified using `struct sockaddr_in`, whereas an IPv6 address is specified using `struct sockaddr_in6`. But `bind()` needs to accept any kind of address, since it works with any kind of socket, so it takes a generic type called `struct sockaddr` and then requires you to cast everything to fit. It's all a bit weird, but for the most part, you can just grab some existing code, so let's get it done. First, we'll set up the basic bits:

```
struct sockaddr_in addr;
bzero(&addr, sizeof(addr));
addr.sin_len = sizeof(addr);
addr.sin_family = AF_INET;
addr.sin_addr.s_addr = INADDR_ANY;
```

Taking it quickly, this declares the address `struct` and then zeroes it out so we don't have to worry about filling every field. It fills out the length field of the `struct` with its true length (because the different types of addresses have different sizes, the receiver needs to know just how much memory is here) and the `family` field with the socket type. Last, we tell it that we just want any address, which here means that it should bind to every interface.

Next, we'll actually bind to a port:

```
addr.sin_port = htons(0);
bind(_socket, (struct sockaddr *)&addr, sizeof(addr));
```

By assigning 0 to `sin_port`, `bind()` knows to just pick a random available port. We don't care what we get, just as long as it's open.

You might be curious about that call to `htons()`. Remember the discussion about endianness? This is where that idea comes in. The "hotns" name actually stands for "host to network short." In other words, it takes a short value and translates it from host (native) endianness to network byte order. This is necessary because the `sin_port` field is specified to be in network byte order, not host byte order. Strange but true!

The next thing is to advertise this service that we've just created. Before we can do that, we need to know what port we're listening on in order to tell the network where to send data. The `getsockname()` call will do that:

```
socklen_t len = sizeof(addr);
getsockname(_socket, (struct sockaddr *)&addr, &len);
```

And then we can make the `NSNetService` to broadcast our presence:

```
// advertise our newly created socket
_advertisingService = [[NSNetService alloc]
```

```
                     initWithDomain:@"local."
                     type:@"_spherenet.udp."
                     name:[[UIDevice currentDevice] uniqueIdentifier]
                     port: ntohs(addr.sin_port)];
          [_advertisingService publish];
```

Note the use of the ntohs() call. It does the opposite of htons(): it takes a short in network byte order and translates it back to host byte order.

We'll also start looking for other services by creating an NSNetServiceBrowser:

```
          // start looking for other services on the network
          _browser = [[NSNetServiceBrowser alloc] init];
          [_browser setDelegate:self];
          [_browser searchForServicesOfType:@"_spherenet._udp."
                     inDomain:@""];
```

And finally, spawn the listener thread and close out the initializer:

```
          // start the listener thread
          [NSThread detachNewThreadSelector:@selector(listenThread)
                              toTarget:self
                              withObject:nil];

      }
      return self;
}
```

The -listenThread method will read all of the data coming in from the network, and we'll
actually write that method a bit later on, after we're finished with the data transmission
code. Let's not forget the dealloc. Even though this object will stick around for the lifetime
of the application, adding dealloc is a good habit:

```
- (void)dealloc
{
      [_advertisingService stop];
      [_browser stop];

      [_advertisingService release];
      [_browser release];

      [_services release];

      [super dealloc];
}
```

Next, we'll implement some NSNetServicesBrowser delegate methods so that we can find
out the results of the search. There are a bunch of delegate methods, but only two matter for

this program: the one that notifies of a new service and the one that notifies that an existing service has gone away.

When a new service appears, it hasn't yet been resolved. In other words, the system knows that the service exists but doesn't know what its address is. We have to explicitly ask the system to resolve the address of any new service that it finds. In addition to that, we'll also add it to our list of services that we want to talk to. When the service goes away, we'll take it back out—nice and easy. We'll also make one more refinement: we don't want to send to our own service, as that just creates confusion, so if the service name matches the current device ID, we're going to ignore that service entirely:

```
- (void)netServiceBrowser:(NSNetServiceBrowser *)browser
                  didFindService:(NSNetService *)service
                  moreComing:(BOOL)moreComing
{
    if([[service name] isEqualToString:
                  [[UIDevice currentDevice] uniqueIdentifier]])
        return;

    [service resolve];
    [_services addObject:service];
}

- (void)netServiceBrowser:(NSNetServiceBrowser *)browser
                  didRemoveService:(NSNetService *)service
                  moreComing:(BOOL)moreComing
{
    [_services removeObject:service];
}
```

Next, we'll implement a method to send a position update to all known services. We'll take advantage of the packet format structs we developed earlier and use them to generate the data to send. The first thing is to declare the appropriate struct and fill out the header:

```
- (void)sendUpdates
{
    PositionPacket packet;
    packet.header.identifier = CFSwapInt32HostToBig(
                              kSphereNetPacketIdentifier);
    packet.header.datatype = CFSwapInt32HostToBig(
                              kSphereNetPositionPacketType);
```

As you may have guessed, the calls to CFSwapInt32HostToBig() are due to endianness again. This time we're using a Core Foundation function to do the byte swapping instead of the POSIX functions from before. This is more a matter of aesthetics than anything, but I prefer the Core Foundation functions here, since they explicitly call out the size of the data being swapped.

Now that the header is complete, we'll fill out the rest of the packet in a similar fashion:

```
packet.r = round(_lastSentUpdate.r * 255.0);
packet.g = round(_lastSentUpdate.g * 255.0);
packet.b = round(_lastSentUpdate.b * 255.0);
packet.x = CFSwapInt32HostToBig(round(_lastSentUpdate.position.x));
packet.y = CFSwapInt32HostToBig(round(_lastSentUpdate.position.y));
```

Next, we're going to send the packet. This is done using the socket function sendto(). It takes a socket, data, options, and an address and sends the data to the given address. No permanent connection is established, and everything is done in a single call, just what we need. The address can be obtained directly from the NSNetService objects. A simple nested loop walks through all the addresses of all the services we've found so far and sends the packet to each one:

```
for(NSNetService *service in _services)
    for(NSData *address in [service addresses])
        sendto(_socket,
                &packet,
                sizeof(packet),
                0,
                [address bytes],
                [address length]);
```

After that, we'll use a delayed perform to schedule this method to run each second if nothing else is going on. By cancelling previous perform requests, we ensure that only one delayed perform is queued at any time:

```
[NSObject cancelPreviousPerformRequestsWithTarget:self
                                         selector:_cmd
                                           object:nil];
[self performSelector:_cmd withObject:nil afterDelay:1.0];
}
```

There is a bit of a tricky shortcut here that merits further discussion, namely the use of _cmd. This is a hidden parameter to every Objective-C method, much like self. The self parameter is implicitly passed into every method and indicates the object that the message was sent to. The _cmd parameter is also implicitly passed into every method and holds the selector that was used to invoke the current method. In this case, _cmd just serves as a shortcut for @selector(sendUpdates), since that's the name of the method that we're using it in.

With this method in place, the implementation for the -localSphereDidMove: method is straightforward. Update the _lastSentUpdate instance variable and then call -sendUpdates:

```
- (void)localSphereDidMove:(SphereNetSphere *)sphere
{
    _lastSentUpdate.r = [sphere r];
    _lastSentUpdate.g = [sphere g];
    _lastSentUpdate.b = [sphere b];
    _lastSentUpdate.position = [sphere position];

    [self sendUpdates];
}
```

That covers sending; now, let's move on to receiving. This is done in a thread where we just loop and try to receive messages from the socket. To receive, we have to allocate a buffer and then call the recvfrom() method, which is the companion to the sendto() method. Where sendto() transmitted some data to a particular address, recvfrom() receives data into a buffer and returns the address that it came from. Returning the address is useful because we need to know the address the data came from to figure out which sphere to update or whether to add a new one. Here's the first part of the loop:

```
- (void)listenThread
{
    while(1)
    {
        PositionPacket packet;
        struct sockaddr addr;
        socklen_t socklen = sizeof(addr);
        ssize_t len = recvfrom(_socket,
                               &packet,
                               sizeof(packet),
                               0,
                               &addr,
                               &socklen);
        if(len != sizeof(packet))
            continue;
```

Note the length check at the end. In network programming, it's important to be extremely paranoid about any data coming in from the network. In this case, if the length isn't what we expected, we'll just ignore the packet entirely.

Next, check the packet header, and make sure the identifier and type are exactly as expected:

```
        if(CFSwapInt32BigToHost(packet.header.identifier)
                != kSphereNetPacketIdentifier)
```

```
            continue;
        if(CFSwapInt32BigToHost(packet.header.datatype)
              != kSphereNetPositionPacketType)
            continue;
```

This is all nice and simple, since we only take one datatype. If we took more than one, the length check would get a bit more complex, and we'd have to take different actions based on the datatype field. Since we're accepting only one datatype, we can simply go back to the top of the loop if it isn't that one.

The next thing to do is to notify the delegate. We don't want to notify the delegate directly, though, since we're not on the main thread, and that could cause all kinds of trouble. Instead, we'll bundle up the appropriate data and then bounce the call over to the main thread and call the delegate from there:

```
        NSAutoreleasePool *pool = [[NSAutoreleasePool alloc] init];
        NSData *packetData = [NSData dataWithBytes:&packet
                            length:sizeof(packet)];
        NSData *addressData = [NSData dataWithBytes:&addr length:socklen];
        NSArray *arguments = [NSArray arrayWithObjects:packetData,
                                                    addressData,
                                                    nil];
        SEL mainThreadSEL = @selector(mainThreadReceivedPositionPacket:);
        [self performSelectorOnMainThread:mainThreadSEL
                        withObject:arguments
                        waitUntilDone:YES];
        [pool release];
    }
}
```

The implementation of -mainThreadReceivedPositionPacket: is straightforward. Unpack the arguments, build a SphereNetSphereUpdate from the packet information and then notify the delegate:

```
- (void)mainThreadReceivedPositionPacket:(NSArray *)arguments
{
    // extract the objects from the array created above
    NSData *packetData = [arguments objectAtIndex:0];
    NSData *addressData = [arguments objectAtIndex:1];
    const PositionPacket *packet = [packetData bytes];
    SphereNetSphereUpdate update;

    // ...and update the SphereNetSphereUpdate struct
    update.r = (float)packet->r / 255.0;
    update.g = (float)packet->g / 255.0;
    update.b = (float)packet->b / 255.0;
```

```
    // ...accounting for differences in endianness
    int32_t x = CFSwapInt32BigToHost(packet->x);
    int32_t y = CFSwapInt32BigToHost(packet->y);

    update.position = CGPointMake(x, y);

    [_delegate networkController:self
                    didReceiveUpdate:update
                    fromAddress:addressData];
}
```

That is everything in the network controller. Now all that remains is to integrate this controller into the view controller, and our work will be complete!

Integrating Networking and the GUI

There are three basic capabilities that need to be added to SphereNetViewController to get it up to speed with all of this nifty networking code we just wrote:

- We need to have it actually create an instance of SphereNetNetworkController and notify that instance whenever the user moves the local sphere around.

- We need to keep track of all the remote spheres and move them whenever an update comes in from the network controller.

- We need to remove inactive spheres after they've been idle for a certain amount of time.

To support these three tasks, we'll be adding three instance variables to the class.

```
SphereNetNetworkController *_netController;
NSMutableDictionary *_remoteSpheres;

NSTimer *_idleRemovalTimer;
```

The network controller and idle timer will be set up in -viewDidLoad, just like _localSphere is.

Back in *SphereNetViewController.m*, add the following to the end of -viewDidLoad:

```
if(!_netController)
{
    netController = [[SphereNetNetworkController alloc]
                                initWithDelegate:self];
    [_netController localSphereDidMove:_localSphere];
}
if(!_idleRemovalTimer)
```

```
{
    _idleRemovalTimer = [[NSTimer scheduledTimerWithTimeInterval:5
                    target:self
                    selector:@selector(idleRemoval)
                    userInfo:nil
                    repeats:YES] retain];
}
```

Notifying the network controller when the local sphere moves takes one additional line in -moveLocalSphereFromTouch:, which now looks like this:

```
if(touch)
{
    [_localSphere setPosition:[touch locationInView:[self view]]];
    [_netController localSphereDidMove:_localSphere];
}
```

To respond to updates, we need to implement the SphereNetNetworkController delegate method. To make sure that we update the right sphere, the address of the sender will be used as the key to look up an existing sphere. If it doesn't exist, we'll make a new one and set it up in the list of remote spheres, and we'll set up its CALayer too. If the remote spheres dictionary hasn't been created yet, we'll create that as well. Finally, we'll set the color and position of the sphere. Here's the code:

```
- (void)networkController:(SphereNetNetworkController *)controller
        didReceiveUpdate:(SphereNetSphereUpdate)update
            fromAddress:(NSData *)address
{
    SphereNetSphere *sphere = [_remoteSpheres objectForKey:address];
    if(!sphere)
    {
        sphere = [[[SphereNetSphere alloc] init] autorelease];
        if(!_remoteSpheres)
            _remoteSpheres = [[NSMutableDictionary alloc] init];
        [_remoteSpheres setObject:sphere forKey:address];
        [[[self view] layer] addSublayer:[sphere layer]];
    }

    [sphere setColorR:update.r g:update.g b:update.b];
    [sphere setPosition:update.position];
}
```

For the final task, removal of idle spheres, the goal is to remove any spheres that haven't been updated in the past 10 seconds. Since there's no explicit disconnection command, this kind of time-out mechanism is the only way for us to detect that a remote copy of SphereNet has gone away.

I've left out an explicit disconnection command more for simplicity than anything else. It simply isn't needed, though adding one might be nice so that disconnected spheres would disappear immediately. Note, however, that even with such a command, this idle removal would still be needed, for the simple reason that the disconnection command isn't guaranteed to arrive. When working with networking code, the fundamental unreliability of networks always has to be considered, and this is no exception.

The idle removal method is nice and straightforward. We simply iterate through the remote spheres dictionary and check each sphere's last update time, a property we cleverly defined at the beginning for exactly this purpose. If the last update time is more than 10 seconds in the past, we remove the sphere from the dictionary and remove its layer from the screen. There's a bit of a dance involved because NSMutableDictionary really dislikes being mutated while it's being iterated, so we store all the dead addresses in an array and remove them all after the fact. Here's the code:

```
- (void)idleRemoval
{
    NSTimeInterval now = [NSDate timeIntervalSinceReferenceDate];
    NSMutableArray *toRemove = [NSMutableArray array];
    for(NSData *address in _remoteSpheres)
    {
        SphereNetSphere *sphere = [_remoteSpheres objectForKey:address];
        if(now - [sphere lastUpdate] > 10)
        {
            [toRemove addObject:address];
            [[sphere layer] removeFromSuperlayer];
        }
    }
    [_remoteSpheres removeObjectsForKeys:toRemove];
}
```

This could potentially get inefficient if there are thousands of remote machines on the network, but in practice, the number of machines is unlikely to ever exceed more than a couple of dozen, so this sort of brute-force search works out just fine.

That's everything! All the pieces are in place now (take a look back at Figure 2-1 to see the completed game). You can start up SphereNet on a bunch of different computers, and everything should be synchronized. If you don't have a bunch of different computers, running one copy on your iPhone and one copy in the simulator is enough to see the network code at work.

If you're looking to build up your experience with networking, the completed SphereNet example offers a good platform to experiment with more sophisticated code. Here are a few ideas for enhancements that you could make:

- Expand the current message format to allow for different shapes and different sizes.

- Add a disconnection message so that disconnected spheres can be removed immediately instead of waiting for a time-out.

- Make updates more efficient by sending only position, not color, data. Send the color separately, in a one-time introduction message. Don't forget to make the protocol detect the introduction and resend it if it gets lost in transit!

Summary

The SphereNet example is now fully functional, and that brings this chapter to a close. First, we developed a basic local version of the program, with an eye toward networking. Next, we examined exactly what we wanted from the networking support and designed a network protocol to reach those goals. From there, we wrote a network controller that implemented that protocol. Finally, we plugged that network controller into the original local code to build the final networked product, and I offered some ideas for extending the application.

The iPhone's great strength is its always-available unlimited network connection. By integrating networking into your application, you can make your iPhone application better by giving your customers the ability to connect with each other from anywhere.

Gary Bennett

Company: xcelMe.com

Location: Scottsdale, Arizona

Former life as a developer: Ten years in the United States Navy aboard two submarines as a nuclear power engineer. Ten years developing Windows, Linux, and Mac OS X applications. Successful IPO in 2002 as the chief information officer of a national health care company. Developing with Windows Visual C/C++, Linux C/C++, Objective-C, MYSQL, and Oracle DB.

Life as an iPhone Developer: EA Sports Tee Shot Live, Colorado Snow Report, Utah Snow Report, RSS Parsing, and SQLite 3

What's in this chapter: This chapter explores processes, threads, race conditions, critical sections, asynchronous programming, deadlocks, threading basics, creating threads, threading dangers, building threading applications, and run loops.

Key technologies

- *Threading*

- *Critical sections*

- *Deadlocks*

Chapter 3

Doing Several Things at Once: Performance Enhancements with Threading

uring my sophomore year in high school, our class got its first computer, an Apple II+. Our teacher opened the top of the computer, and we all looked in and marveled at the 6502 processor and the amazing amount of memory, 48KB of RAM. That is when I became hooked on technology, specifically computer science.

After high school, I joined the Navy and was a nuclear engineer aboard two different submarines, a Los Angeles–class, fast-attack submarine and an Ohio-class ballistic missile submarine. While serving at my last duty station in Bangor, Washington, I heard a university had opened a satellite campus on base and was offering night classes leading to a four-year degree in computer science. I couldn't sign up fast enough. It was 1990, and most of my instructors were either working at Microsoft on a new operating system technology or military contractors who worked on really cool classified stuff. The Microsoft employees said this "new technology" they were working on enabled true preemptive multitasking and the ability to thread applications.

Our little satellite campus didn't have access to big expensive Unix computers. Microsoft wasn't about to let us pore through their new technology under development, but we did have Alfred V. Aho and Jeffrey D. Ullman's infamous Dragon book (that is, *Principles of Compiler Design*), DOS, and a Unix-like operating system called MINIX, which inspired Linus Torvalds to develop the Linux operating system years later. We spent many hours poring over the code with our instructors learning how it related to the new technology. I rapidly saw the power in multitasking and threading and how it can improve the user experience.

Within months of the Windows NT release, I was out of the Navy with my degree and, more importantly, a background in Windows NT and writing multithreaded applications. Over the years, I moved from writing Windows applications to Linux applications. In 2002, I moved to Mac OS X. Then in 2008, I heard that the iPhone was going to have Mac OS X as its operating system, and once again, I couldn't sign up fast enough. The iPhone was going to have the ability to support various types of threading APIs: POSIX Threads, NSObject, NSThread, and NSOperation.

Even as a Windows and Linux C/C++ programmer with a background in Objective-C and Cocoa, I found the learning curve steep on my first iPhone application. I also found the process of getting my first iPhone application to actually run on the iPhone very difficult.

Having said that, I am amazed at the power, features, and maturity of the iPhone SDK. A year after the iPhone SDK came out, developers had access to tools, the SDK, documentation, a simulator, and a software distribution system unparalleled in the history of software development. All this is due, in part, to the powerful Mac OS X operating system that the iPhone operating system is based on. The iPhone operating system is a true preemptive, multitasking operating system that has the ability for the developer to thread applications.

Other authors in this book have already pointed out that threading your iPhone applications sometimes introduces another level of complexity and should be avoided if possible. No one sets out to write a threaded application. However, threading your application is often necessary to accomplish your desired effect. Many applications on your computer are threaded. Many iPhone applications are threaded and, if properly designed, can be maintained. Like anything else, everything starts with the creator's understanding of the tools and how to properly use them.

Threading enables computer programmers to have their software accomplish multiple tasks simultaneously. There are times when your application may need to work for several seconds and you don't want the user to wait while a task is being accomplished. For example, your application may need to download information from the Internet. You may want the user to be able see an activity indicator on the screen or be able to navigate to another part of the application while the download occurs in the background. Threading enables you to add this type of responsiveness to your application.

This chapter will cover the main principles of threading. We will develop a neat little iPhone application that demonstrates all these principles. Additionally, we will consider common pitfalls associated with threading and how to avoid them.

Beginning to Write Threading Applications

I am simply amazed that, when I am holding the iPhone, I am holding a preemptive, multi-tasking computer with Unix, a graphical interface, GPS functionality, Wi-Fi, a ton of built-in libraries, and by the way, a cell phone. Amazing!

Here are the terms and concepts I am going to break down and describe before we hop into our threaded iPhone application:

- Thread
- Process
- Multitasking
- Synchronization
- Critical section
- Race condition
- Mutex
- Deadlock

When learning new and sometimes intimidating concepts, I tell the students I teach at xcelme.com to learn the concepts and then apply them using KISS—"keep it simple, silly." We are going to do the same thing. After we have learned these threading concepts, I am going to apply them in a basic iPhone application that will enable the user to spawn off multiple threads as well as control their termination and access to data. With this application, you will understand basic threading concepts and be able to decide when to use threading. Most importantly you can use this information to create cool iPhone applications.

Knowing When to Thread

In computer science a **thread**, or **thread of execution**, is a fork in your computer program that produces two or more concurrently running tasks. The thread is contained in the computer program.

Basically, you know you need to thread your application when your application needs to be responsive to the user while completing one or more tasks at the same time. Many times, the iPhone SDK will make it very apparent to you when this needs to occur, which I found out the hard way: the application I was developing for a client needed to read multiple RSS feeds and provide a visual cue to the user that the application was functioning while the user waited. This is normally accomplished by using an activity spinner in the center of the screen. I wanted to dim the background and show the activity spinner at the same time the application was reading and parsing the RSS feed. To do this, I needed to spawn a thread to dim the background and show the activity spinner. Again, I didn't set out to write a threaded application when I started writing the Colorado Snow Report application (see Figure 3-1), but when I looked into how to use the spinner function, I was left with no choice. Fortunately, the iPhone SDK made threading possible without adding much complexity.

People sometimes ask, "Why should I thread; I only have one processor?" Right now, that is correct; today's iPhone currently has only one CPU. That will surely change over time as multiple CPUs and graphics processors are added, and iPhone applications are made available on other Apple touch-screen devices. More importantly, one processor is capable of performing most instructions faster than the user can deliver to it. Like most modern-day operating systems, the OS X operating system is capable of multitasking. **Multitasking** means the operating system will share CPU time with more than one process so that all processes get waited on. The operating system will let the CPU service a process for a few milliseconds and then it requires the process to sleep while the CPU services other processes.

This preemptive multitasking is ideally suited for the iPhone. As your application is running and a phone call comes in, the operating system steps in, puts your application to sleep, services the phone call, and after the phone call is complete, services your application again. In a preemptive, multitasking environment, the operating system prioritizes processes and time-shares as they run, so no process starves by not being serviced.

Figure 3-1. The iPhone Simulator running Colorado Snow Report. Notice the application background is grayed and the activity indicator is spinning while the application fetches data over the Internet.

Understanding Threading Basics

The best way to discuss any new programming concept is to make sure you understand some key words and how they apply.

The first important term is "process." A **process** is simply your application running; Figure 3-2 illustrates a simple process. Every process gets its own address space and call stack to keep track of methods, variables, and events.

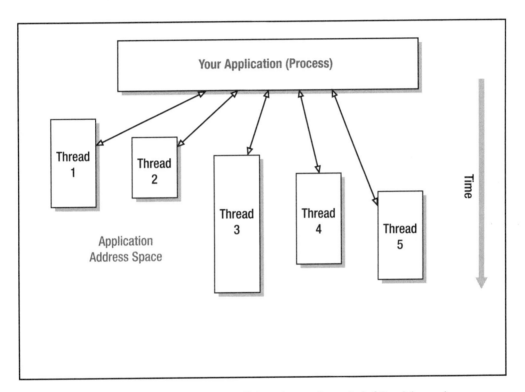

Figure 3-2. *An iPhone application spawning off threads over the period of time it is running*

As an application is running and needs to spawn a thread, it passes to the thread an object for the thread to operate on. The thread may run for a few seconds and terminate itself, or the thread can run in a loop for the life of the process. When the process terminates, so do all threads.

The operating system will give your thread a time slice for the processor to service it. Don't rely on the operating system to give your thread a specific amount of time or priority to your thread. If you spawn four identical threads at the same time, don't expect your threads to finish in the order they were spawned. This introduces the concept of **synchronization**.

If multiple threads try to access the resources for a read and write operation at the same time, the values for the resources might not be accurate and may become corrupt. For example, assume two threads representing two characters in a game application run simultaneously. Synchronization refers to keeping the data coherent or to maintaining data integrity between the threads. If one thread tries to read the gameScore variable while the other tries to update the gameScore variable, the gameScore variable could be inaccurate, as shown in Figure 3-3.

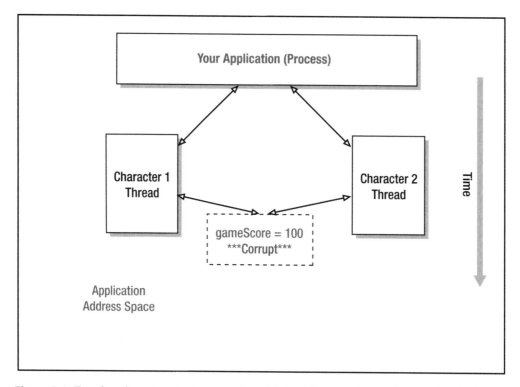

Figure 3-3. *Two threads attempting to access shared data at the same time and corrupting the data*

A **critical section** is a piece of code that protects shared resources from more than one thread accessing that section of code at a time. To enter a critical section, a thread must obtain a semaphore (which gives it permission to enter the critical section), access the resources, and then leave. While the thread is in the critical section, no other threads are allowed to access this section of code, as shown in Figure 3-4.

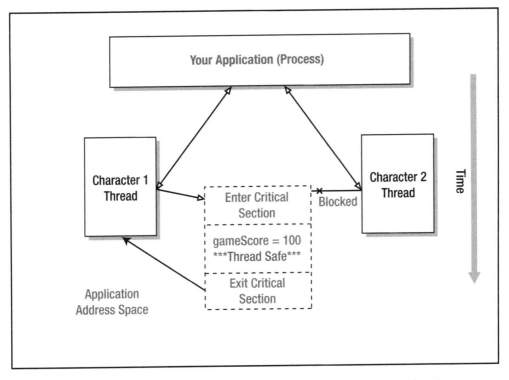

Figure 3-4. *Two threads successfully accessing data at the same time using a critical section*

Avoiding Threading Pitfalls

This chapter will focus on one of the common concerns in multithreaded applications—shared data. However, you should be aware of many concerns when writing threaded applications.

One common pitfall of threaded applications occurs when the application unexpectedly becomes dependent on the timing of the threads' execution. When certain conditions exist, the timing of the threads' access to resources can causes unexpected results. This situation is usually referred to as a **race condition**. For example, your application may spawn several threads to access the Internet. Nine times out of ten, your application works perfectly because the third thread finishes its task last. However, when thread three finishes first, your application crashes, because you never expected that to happen. A solution can be for each thread to lock all the resources it needs before beginning its task. This is accomplished through the use of **mutexes** ("mutex" is short for "mutual exclusion," which means one or the other but not both). Mutexes are used in threading to avoid the simultaneous use of a common resource. Several famous examples of race conditions have caused disastrous effects. One was the near loss of NASA's Mars Exploration Rover, *Spirit*, shortly after it landed. Another race condition resulted when several faults occurred in an energy management system provided by GE Energy and caused an unexpected thread order, which lead to the power blackout in North America in 2003.

Another common pitfall in threaded applications occurs when two or more threads are blocked forever, waiting for each other to release a resource they have reserved independently. This is called a **deadlock condition**.

Figure 3-5 shows a deadlock condition occurring when two threads are each waiting for the other to release its database table lock before it can continue. Character 1's thread locked the score table and then the shield table. Character 2's thread locked the shield database table and then the score database table. Neither thread will be able to continue, as a deadlock condition has occurred. You will see in our example application how this can occur and how to prevent it.

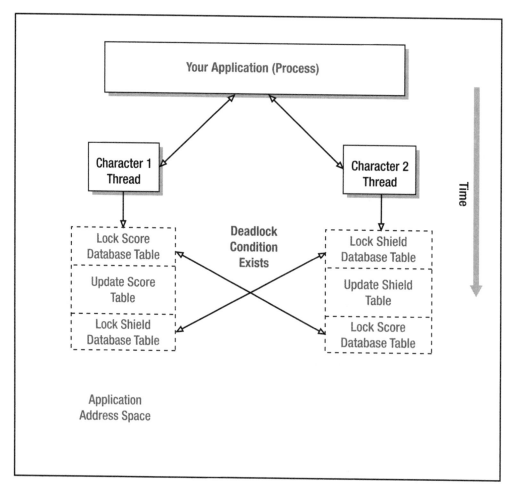

Figure 3-5. *Two threads causing a deadlock condition*

Writing the Thread the Needle Application

Enough explanations, let's write our application. In this section, we are going to write our threaded application. We will connect our outlets and actions and then write our threading code.

Building Our Application

This simple iPhone application example demonstrates how to overcome one of the most common issues developers have with threading applications—how to share data between threads.

Let's write the simple iPhone application shown in Figure 3-6, so that we can focus on the specifics of threading. The application enables the user to spawn up to four threads. The user can kill any thread at any time or kill all threads at the same time. All four threads will access the same shared variable. We'll call the application Thread the Needle.

Figure 3-6. *The iPhone Simulator running our Thread the Needle application with all threads processing thread increment counts*

1. To begin, open Xcode, and create a new project.

2. Select View-based Application, as shown in Figure 3-7.

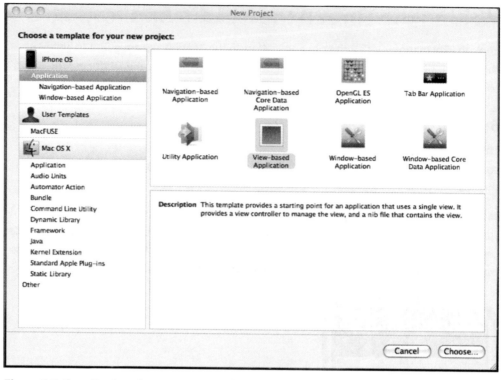

Figure 3-7. *Open Xcode and create a new view-based application.*

3. When Xcode asks to name your project, name it **Threading** (we will keep the name short for our example), as shown in Figure 3-8.

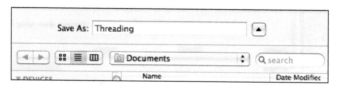

Figure 3-8. *Save your project as Threading.*

4. This will create a project with two Interface Builder XIB files, named *MainWindow.xib* and *RootViewController.xib*, and the *Threading-Info.plist* where application-launching metadata is stored.

5. Once we have created our project, we need to open Interface Builder to lay out our iPhone application. Double-click the *ThreadingViewController.xib* file to open it in Interface Builder (see Figure 3-9). Once Interface Builder opens the XIB file, double-click the View icon to open the layout for our view.

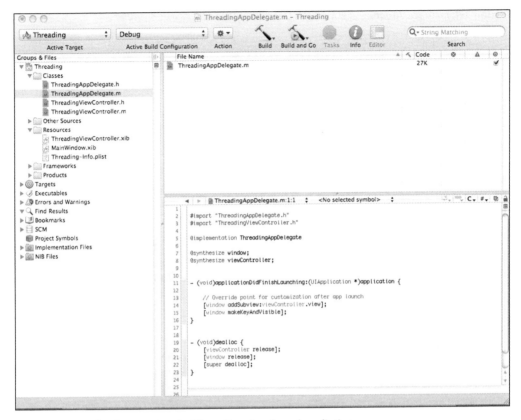

Figure 3-9. *Xcode after creating the Threading view-based appliation*

6. Now, select **Tools ➤ Library** from the menu. In the Library window, select **Inputs & Values**. We are ready to lay out our view.

7. Drag and drop a Label and a Round Rect Button object into your view and make the view look like Figure 3-10. You can double-click the objects to change the titles.

Next, we need to add our instance variables and methods to hold the values and perform the actions from the view we just made. Let's go back to Xcode and open our interface file that describes the ThreadingViewController class. Open *ThreadingViewController.h*, and modify the code to look like Listing 3-1.

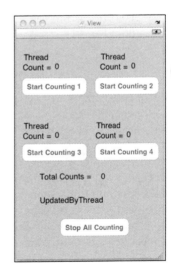

Figure 3-10. *Make your view layout look like this.*

Listing 3-1. *Instance Variables and Methods Describing ThreadingViewController*

```
#import <UIKit/UIKit.h>
@interface ThreadingViewController : UIViewController {
    bool button1On;// keeps track if the Start Counting buttons are clicked
    bool button2On;
    bool button3On;
    bool button4On;

    int total;  // keeps track of the total count
    int countThread1; //keeps track of the count for each thread
    int countThread2;
    int countThread3;
    int countThread4;

    NSLock *myLock; //mutex used to create our Critical Section

    IBOutlet UILabel *thread1Label; //thread value labels
    IBOutlet UILabel *thread2Label;
    IBOutlet UILabel *thread3Label;
    IBOutlet UILabel *thread4Label;

    IBOutlet UILabel *totalCount; //total thread count label
    IBOutlet UILabel *updatedByThread; //updated by thread label

    IBOutlet UIButton *button1Title; // buttons titles to be updated when
      //clicked
    IBOutlet UIButton *button2Title;
    IBOutlet UIButton *button3Title;
    IBOutlet UIButton *button4Title;
}

@property (retain,nonatomic)  UIButton *button1Title; //getter and setters
@property (retain,nonatomic)  UIButton *button2Title;
@property (retain,nonatomic)  UIButton *button3Title;
@property (retain,nonatomic)  UIButton *button4Title;

@property (retain,nonatomic) UILabel *totalCount; //getter and setters
@property (retain,nonatomic) UILabel *thread1Label;
@property (retain,nonatomic) UILabel *thread2Label;
@property (retain,nonatomic) UILabel *thread3Label;
@property (retain,nonatomic) UILabel *thread4Label;
@property (retain,nonatomic) UILabel *updatedByThread;
```

```
-(IBAction)launchThread1:(id)sender; //methods each button will trigger
 //when clicked
-(IBAction)launchThread2:(id)sender;
-(IBAction)launchThread3:(id)sender;
-(IBAction)launchThread4:(id)sender;
-(IBAction)KillAllThreads:(id)sender;
```

After you save the file, we're ready to connect our outlets and actions in Interface Builder. Double-click *ThreadingViewController.xib*, and right-click the File's Owner icon. Connect your outlets as shown in Figure 3-11. Drag from the Outlets section to the appropriate Start Counting button.

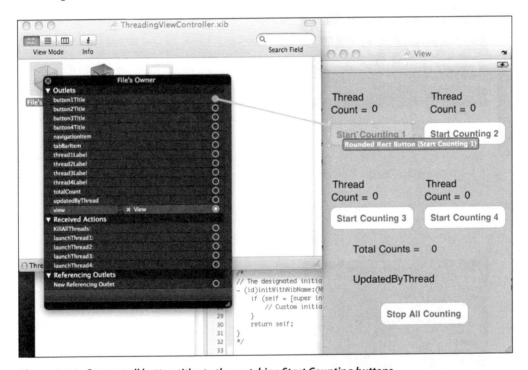

Figure 3-11. *Connect all button titles to the matching Start Counting buttons.*

Now, connect the thread labels to the outlets, as shown in Figure 3-12. Drag from the Outlets section to the corresponding Thread Count label.

Now, we need to update the total increment counts for all threads by dragging the `totalCount` outlet to the Total Counts label; see Figure 3-13. This label displays the total of all thread increment counts. Each thread, in increments of ten, updates this value.

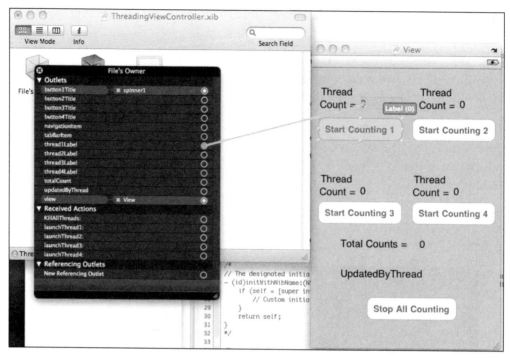

Figure 3-12. *Connect all thread labels to the corresponding Thread Count values.*

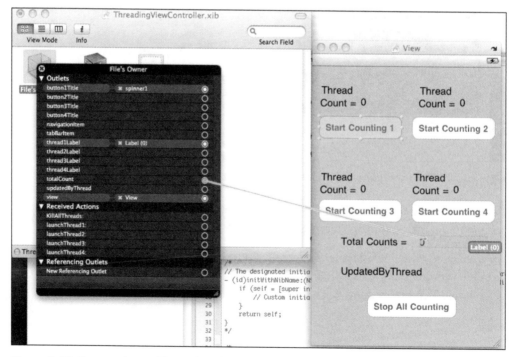

Figure 3-13. *Connect the totalCount thread to the associated label.*

Next, we will update the updatedByThread outlet to the associated label, as shown in Figure 3-14. This label lists the last thread to update the Total Counts value.

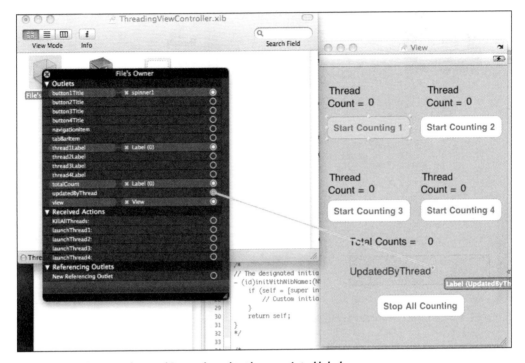

Figure 3-14. *Connect the totalCount thread to the associated label.*

Finally, we will connect all the buttons with their associated events, as shown in Figure 3-15. Control-click the button and drag to the File's Owner icon. Then, select the associated event for each button to enable the button to trigger an event each time it's clicked. The event listener will then call the associated method for that event. Repeat connecting the actions and outlets for the remaining buttons and labels.

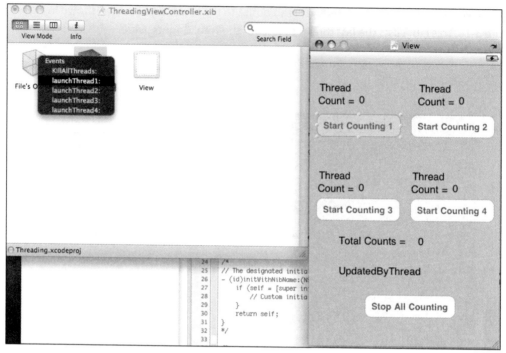

Figure 3-15. *Connect the events to the associated buttons.*

Creating a Thread

The interface section is now described. Save the changes in Interface Builder, and let's focus on the implementation section of our ThreadingViewController. Open the *ThreadingViewController.m* file in Xcode.

We need to complete implementing our getter and setter methods. To do so, add the code in Listing 3-2 to the implementation section of the *ThreadingViewController.m* file.

Listing 3-2. *Completing the Getter and Setter Methods in ThreadingViewController.m*

```
@synthesize totalCount;
@synthesize thread1Label;
@synthesize thread2Label;
@synthesize thread3Label;
@synthesize thread4Label;
@synthesize button1Title;
@synthesize button2Title;
@synthesize button3Title;
@synthesize button4Title;
@synthesize updatedByThread;
```

We need to look at how to implement our thread when a user wants to launch a thread by clicking a button. Clicking a Start Counting button will trigger an event that will be handled by the corresponding launchThreadX methods.

Creating a Worker Thread

It is important now to discuss how we are going to implement a worker thread and what the thread is going to do. When the thread is spawned, it increments a counter ten times, updates its display on the iPhone, sleeps after each increment, and updates the total thread counter at the end of each cycle. The cycle repeats until the user signals to kill the individual thread or kill all threads (by clicking Stop All Counting). See Figure 3-16.

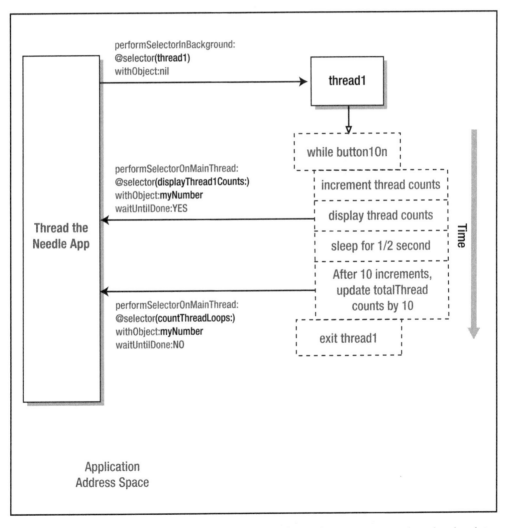

Figure 3-16. *Diagram showing when the Thread the Needle application spawns a thread and updates the view display*

With OS 10.5 (Leopard), NSObject gained an additional method called performSelectorInBackground: withObject:. This method makes it great for developers to spawn a new thread with the selector and arguments' provided. You will notice, if you look at the threading APIs, that this method is just about the same as NSThread class method + detachNewThreadSelector with the added benefit that with the NSObject method, you no longer have to specify a target. Instead, you are calling the method on the intended target.

When you call performSelectorInBackground: withObject:, you are essentially spawning a new thread that immediately starts executing the method in your class. This convenience method puts your application into multithreaded mode. Let's implement the code when a user clicks a Start Counting button; see Listing 3-3. Go ahead and implement the code for the remaining Start Counting buttons as well.

Listing 3-3. *The launchThread1:(id)sender Method*

```
-(IBAction)launchThread1:(id)sender
{
    if (!button1On)
    {
        button1On = TRUE;
        [button1Title setTitle:@"Kill Counting 1"
forState:UIControlStateNormal];
        [self performSelectorInBackground:@selector(thread1)
withObject:nil];
    }
    else
    {
        button1On = FALSE;
        [button1Title setTitle:@"Start Counting 1"
forState:UIControlStateNormal];

    }
}
```

In Listing 3-3, we first check to see what state the button is in and then we change it. Next, we change the title of the button to reflect the state. If the button is being clicked for the first time, we will launch our thread (see Listing 3-4). Implement the remaining thread logic for the other three buttons.

Listing 3-4. *Launching thread1*

```
-(void)thread1
{
    NSAutoreleasePool *apool = [[NSAutoreleasePool alloc] init];
    NSNumber *myNumber = [NSNumber numberWithInteger:1];
```

```
    while(button1On)
    {
        for (int x=0; x<10; x++)
        {
            [self performSelectorOnMainThread:@selector
                                        (displayThread1Counts:)
                                        withObject:myNumber
                                        waitUntilDone:YES];
            [NSThread sleepForTimeInterval:0.5];
        }
        [self performSelectorOnMainThread:@selector(countThreadLoops:)
                                        withObject:myNumber
                                        waitUntilDone:NO];

    }
    [apool release];
}
```

The first thing you should notice in Listing 3-4 is that we are responsible for the memory pool. That is correct! When we launch a thread, we are essentially leaving the Cocoa framework. When we do that, we are responsible for cleaning up the memory pool. If we don't, we will leak memory.

We then begin a simple loop. We want to keep our worker threads busy doing something in this example, and this simple loop is an easy way to do that without creating a run loop. We will discuss run loops shortly.

Our thread increments and then we see a new method call, performSelectorOnMainThread.

Apple warns against calling the main thread from a worker thread, [self displayThread1Counts:], explaining that unexpected results may occur. Apple is right; I have tried it. So, to call our displayThread1Counts: method, we need to launch a thread by setting waitUntilDone to YES. This tells my current worker thread to wait until Thread1Counts is done before continuing. I did this to illustrate the waitUntilDone. I wanted to slow down the thread at this point, so I put the thread to sleep for half a second. After the loop completes, we now want to update the Total Count field on the iPhone. We now launch another thread to call the countThreadLoops: method.

This time, we don't wait for the method to finish. We spawn the thread and let it operate asynchronously. Asynchronous events are those occurring independent of the calling thread or main program flow. **Asynchronous** just means the calling thread doesn't sit and wait for the response.

Creating an Event Processing Loop

When we click a Start Counting button in our iPhone application, we can spawn a thread with performSelectorInBackground: withObject. Now that our worker thread is working we need to update the Thread Count field for each thread. To do that, we have made a method to do that, called -(void)displayThread1Counts:(NSNumber*)threadNumber (see Listing 3-5).

Listing 3-5. *Updating Our Thread Counts*

```
-(void)displayThread1Counts:(NSNumber*)threadNumber
{
    countThread1 += 1;
    [thread1Label setText:[NSString stringWithFormat:@"%i", countThread1]];

}
```

As mentioned earlier, we want to keep our worker threads busy. I have implemented a simple loop to this with a sentinel variable, button1On, to tell the thread when to exit. This works fine for this example, but there may be times when you need more granularity in communicating with your threads. To accomplish this, Apple provides run loops as part of the infrastructure associated with all threads. A **run loop** processes the events that you use to schedule work and coordinate the receipt of incoming events. Its purpose is to keep your threads busy when there is work to do and put your threads to sleep when there's none. For more information on run loops, see Apple's "Threading Programming Guide."

Implementing a Critical Section

Next, we need to implement the critical section in Listing 3-6 to protect our shared variables.

Listing 3-6. *Implementing Our Critical Section with NSLock*

```
-(void)countThreadLoops:(NSNumber*)threadNumber
{
    [myLock lock]; //mutex to protect critical section
    total += 10;
    [self.totalCount setText:[NSString stringWithFormat:@"%i", total]];
    [self.updatedByThread setText:[NSString stringWithFormat:@"Last updated
                by thread # %i",[threadNumber integerValue]]];
    [myLock unlock]; //make sure you perform unlock or you will create a
      //deadlock condition
}
```

First, we lock our critical section with an instance of NSLock. As described in Figure 3-4, this mutex prevents other threads from accessing this section of code when a thread is already in there, preventing unexpected results. We then update the total number of thread counts by ten, and update our view with this count and the thread that sent the message.

Last, but certainly not least, we need to unlock our critical section; otherwise, all threads will be blocked from entering this section of code, and we'll lock up our application (recall that this condition is called a deadlock).

Stopping Multiple Threads at Once

The final thing we need to implement is the ability to stop all threads at once. I implemented a simple method called -(IBAction)KillAllThreads:(id)sender;. This method simply sets all of our buttonxOn to FALSE (see Listing 3-7). All threads that are in their while loop will exit these loops, and the threads will gracefully terminate after the loop has been completed ten times. You may notice a little delay from the time you click the button until the threads are actually done counting. The looping occurring ten times causes the delay (see Listing 3-4).

This rudimentary signal is sufficient for our application. If you need greater granularity, you should implement a run loop for your thread, as discussed previously.

Listing 3-7. *A Simple Signal to Gracefully Stop All Threads*

```
-(IBAction)KillAllThreads:(id)sender
{
    button1On = FALSE;
    button2On = FALSE;
    button3On = FALSE;
    button4On = FALSE;

    [button1Title setTitle:@"Start Counting 1"
forState:UIControlStateNormal];
    [button2Title setTitle:@"Start Counting 2"
forState:UIControlStateNormal];
    [button3Title setTitle:@"Start Counting 3"
forState:UIControlStateNormal];
    [button4Title setTitle:@"Start Counting 4"
forState:UIControlStateNormal];
}
```

Summary

With the Thread the Needle application, we have implemented a very simple and useful example of threading an iPhone application and controlling the threads. You should now understand that threading can give the user more control over the application and make the application more responsive to the user while background work is being performed. A threaded application can add a more professional touch to your application; with it, you can improve application responsiveness and perform tasks in the background independent of the user.

Threading your application does present some dangers, particularly race conditions and deadlocks. However, if you protect shared data with critical sections, as you learned about in this chapter, you can over come these dangers.

This little device is able to unleash an absolutely amazing amount of power when in the hands of a knowledgeable developer. Good luck and have fun!

Matthew "Canis" Rosenfeld

Company: *Wooji Juice*

Location: *London, UK*

Former life as a developer: Programming for a bit more than a quarter of a century, including three years of studying interactive design. Videogame developer for several years at Mucky Foot, working on Urban Chaos (PC and PSX), the BAFTA-nominated StarTopia (PC), and Blade II (Xbox and PS2). Senior programmer at Sony Computer Entertainment Europe for several years. Started Wooji Juice to focus on Mac and iPhone games and serious applications, and released Voluminous (for Mac OS X) and various iPhone applications.

Favorite programming languages are Python and Objective-C. Lots of C++ experience gives me plenty of reasons not to list it as a favorite, though it still comes in useful occasionally. Spent particularly large chunks of my game development career working on networking, scripting and virtual machines, audio engines, editing tools, creature behaviors, and creating user interfaces.

Life as an iPhone developer: Designed and wrote the Stage Hand business application and the Hexterity puzzle game. Lead engineer for the KarmaStar strategy game (contract work for Arkane, published by Majesco).

What's in this chapter: This chapter looks at iPhone's multitouch capabilities and how they're made available, designing touch and multitouch input schemes, handling touch events, recognizing specific gestures (e.g., swipe, pinch/unpinch) and distinguishing them from other similar gestures, and using inertia to give user-interfaces "weight."

Key technologies

- *Multitouch*
- *Gesture detection*
- *Human-interface issues*

Chapter 4

All Fingers and Thumbs: Multitouch Interface Design and Implementation

n this chapter, we'll be examining what is, perhaps, the iPhone's defining feature: touch input. Although many devices have used touch screens over the years, the iPhone was the first to bring multitouch to the masses. It also takes a different approach to input in general. Where many have transferred desktop-style interfaces to touch-screen devices, using a stylus in lieu of a mouse, the iPhone creates a new set of conventions, swapping pointer accuracy for responsiveness, simplicity, and direct manipulation of the interface.

iPhone's touch screen seems effortless to use, because Apple has spent a lot of time working out the subtle details. If you use the built-in controls, you can take advantage of this for free, but if you need to implement your own controls, paying attention to these details makes the difference between an application that is awkward and one that feels natural and iPhone-like.

So in this chapter, we'll look at those details through the lens of Stage Hand, a remote control for Keynote (Apple's slick presentation package) and one of the first applications to appear in the App Store. Stage Hand required a lot of detail work for processing touch input. We'll look at why this is, how the problems were solved, and the general principles that can be applied elsewhere. We'll also put together a simple application that demonstrates a variety of touch gestures and the programming techniques behind them.

Looking at the iPhone's Capabilities

In 2007, stuck at work late on a cold winter's evening, I was following the live blog of a Ste-venote in a window on my screen—*the* Stevenote: you know the one, where the Coming of the Jesusphone was foretold.

I don't generally get excited by these events. I'm curious to see the new toys, but I guess I don't have the same holy fervor as some of my brethren. But this one was different. While the Apple phone rumors had been swirling for a few days, they'd truly been swirling for *years* to little apparent effect, and it seemed likely that, even if the rumors proved true, the iPhone would be an incremental upgrade to the iPod.

But no, this really *was* different. The magic words on screen were "Runs OS X." The device was practically a Mac in your hand, with some familiar frameworks to boot, like Cocoa and Core Animation. We knew of these frameworks: we had tamed them, and here they were, being dangled tantalizingly in a svelte new outfit.

Three things became immediately obvious:

- I wanted this device.

- I wanted to code for this device.

- For once, it looked like it actually *was* going to change everything.

Later that night, I walked with a colleague through the rain and stench on the streets of Lon-don's Soho in search of a good noodle place with indoor seating to spare. We threw crazy ideas around for applications and speculated about what the iPhone actually held in store.

As is often the case with Apple events, there was as much left unanswered as shown. At this point, it wasn't even clear whether third-party applications would be allowed, or if the platform would remain forever closed. We wouldn't get all our answers for more than a year, with the release of the iPhone SDK.

One thing that worried us was that, by the time the SDK was available, the MacBook Air had already been released, with its own variety of multitouch and an API exposed to program-mers. And that API was limited: the OS would determine pinch, swipe, and rotate gestures and translate them into mouse scroll and zoom events. An application wasn't able to invent its own gestures or process multiple-finger touches independently. Would the iPhone be the same?

Fortunately, the iPhone multitouch system is actually quite thoroughly exposed to the programmer. It can track up to five simultaneous touches—conveniently a whole hand's worth—and it does a lot of processing behind the scenes to give you fairly clean data to work with. The hardware appears to pick up contact patches: think about the oval shape someone's nose makes when pressed against a window, the effect is similar when your finger taps the screen. Then, it identifies a single point within that oval that represents the touch point. It takes care of figuring out the difference between one big finger and two small ones smooshed together. It also tracks each touch independently, so you can figure out what a finger does over the lifetime of a given period of contact with the screen, rather than the OS just throwing a bucket of coordinates at you.

However, Cocoa Touch doesn't give you any interpretation of events; early revisions of the SDK performed detection of certain actions, such as swipes, but these were unceremoniously yanked. You need to figure out what touches mean for yourself or rely on existing controls (e.g., UIScrollView) that handle the touches but often have limited features.

As well as the technical side of things, there are a number of other design issues to consider. You still have to bear in mind that the contact patch represents quite a large area; just place your fingers against an iPhone, and you'll see that a finger can be a quarter the width of the screen, maybe more.

Apple recommends that controls have an active area of at least 44 pixels square, but it's important to remember this need not match the on-screen graphic. In fact, if you observe the behavior of the built-in controls, the tap region is often much larger than the control's apparent size. This disparity is both masked by, and in response to, the size of the average finger, but if you try poking at the iPhone Simulator with a mouse, you can see the effect in action. For example, the back arrow button in the top-left corner of many iPhone screens actually extends out over the status bar, to the corner of the screen (see Figure 4-1).

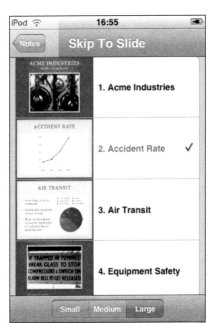

In general, Apple's approach seems to be very generous in the areas you can tap to perform actions, while guarding against accidental taps by automatically filtering out certain kinds of touch (such as mashing the screen with your whole hand). Another common pattern is to use one hit-test region to detect finger-down events and then expand that region for finger-up events. Open the iPhone's Clock application, and switch to the Timer tab. Press *and*

Figure 4-1. The tappable region of the back button, highlighted, extends not only beyond the bounds of the button itself but beyond the navigation bar that contains it and over the status bar.

hold the Start button; you'll see it darken in response. Now, slide your finger off the button, up or down, and see how far you can move it before the button resets. You can slide your finger more than twice the size of the button!

Designing for Multitouch

Strangely, it was the Apple TV that first led Wooji Juice down the road of enhancing Apple's Keynote application. Having spent much time waiting while harried technicians scurried in and out, trying to get a guest presenter's equipment hooked up to boardroom displays, I was thinking that a smarter solution would be to leave an Apple TV hooked up and just stream the presentation across Wi-Fi.

Of course, the Apple TV is a closed system, and besides, the resources for a project like that were out of our reach. But much later, having quit the day job to do iPhone development and needing a project to work on, I remembered that idea. Of course, it still wasn't practical, but musing on Keynote, I remembered how my old phone had a Bluetooth remote-control feature. Ostensibly, you could control presentations with it, but none of the buttons matched up with their effects, and all you could really do was step forward or backward. But not being tied to the keyboard to control your presentation was nice. I wanted something like that for the iPhone.

The iPhone's unique features meant we could go way further than simply stepping back and forth through the presentation. I'd always appreciated Keynote's Presenter Display, and the iPhone's beautiful, crisp screen meant we could place some of that information right there with you. I thought of presenters I'd seen holding stacks of index cards in their hands, shuffling through their notes as they went: we could do this on the phone, but with no possibility of nervously fumbling and scattering cards all over the floor.

One of the things that seemed important was to avoid cluttering the screen with buttons. In particular, we wanted the basic features to be usable without even looking at the screen. Not only would buttons reduce the amount of space for displaying notes, it would make it far more likely that—given the lack of haptic feedback—a presenter might trigger features by accident. A ground rule for our design has been the "principle of least embarrassment." If we had to make the choice, we'd rather require slightly more effort to use a function than have something go embarrassingly wrong during a presentation. This principle has had a definite impact on our user interface and handling of touch events.

Much of Stage Hand's development was spent dealing with the same things every new iPhone developer runs up against, `UIViewControllers`, getting builds running on real hardware, that sort of thing. When we started out, I think Interface Builder didn't yet support iPhone development. The rest of the time was wrangling Keynote, whose scripting API appears to have a number of major bugs and is missing a lot of features that would've been useful.

With just a few days to spare, we got Stage Hand shipped off to Apple and accepted for publishing on the App Store, ready for the big launch on July 11. Perhaps due to time zones, the store opened early from our perspective (we'd assumed it would open on Cupertino time) and so late on the tenth we scrambled to switch over our web site and, like every other application developer, nervously read tea leaves to try to figure out if anyone was buying (at the time, no sales reports were available).

Soon after, feature requests started arriving, and one of the most popular caused a few design headaches. You see, originally Stage Hand was intended to be used just in Notes View mode—portrait orientation, with your Presenter's Notes shown on a panel we call the slate (see Figure 4-2), which you flick from side to side with your finger, much like the iPhone's native Weather application. To support the high-lighter feature, you could tip the phone on its side to see a screenshot of your current slide, so you'd have something to aim with; Highlighter mode is shown in Figure 4-3.

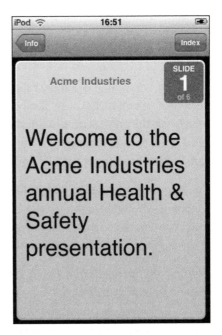

Figure 4-2. *The entire gray slate can be dragged from side to side, or the contents can be flicked up and down.*

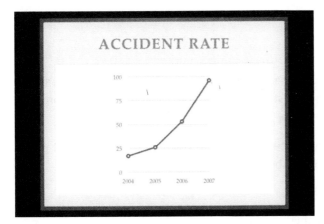

Figure 4-3. *Originally called Highlighter mode, this is now known as Slide View mode, for obvious reasons—at least, in hindsight.*

For many people, especially those who don't use notes, this was the most natural way to view the presentation. To them, this wasn't Highlighter mode, it was Slide View mode, and why on earth couldn't you change slides?

Doh. Well, we wanted to support slide changes. But how? Swiping from side to side didn't seem like an option, because placing your finger on the slide would trigger the highlighter, which was the original purpose of the view after all. Buttons were also out, because they'd reduce the already-limited space for displaying the slide, reducing not only its readability but also the accuracy of the highlighter.

The solution we turned to was multitouch. By using gestures for control, we could add more functions without needing to take up screen space. The downside is that cramming gestures into an interface can make it fiddly, and discerning between gestures can be difficult, which may lead to accidental commands that fail the principle of least embarrassment.

Fortunately, our needs matched quite well with gesture control. Resizing the highlighter, for instance, naturally lends itself to a pinch/unpinch gesture. In fact, we arrange for the highlighter to fit between the user's fingers, wherever the fingers are placed, which offers direct and intuitive control, as well as providing *Kids in the Hall* fans with "I'm crushing your head!" flashbacks as a bonus.

The gestures/events we needed to handle follow:

- Swiping left and right between slides

- Activating and moving the highlighter

- Resizing the highlighter

- Dismissing the highlighter

We also wanted to avoid toggling modes on and off, because in a stressful situation, it's all too easy to toggle one time too many and start flailing around because the application is in the wrong mode.

We actually tried a number of options (some still available from the Settings, for those who prefer them), but the solution we found to be the best is this:

- Single-finger swipe to control slides

- Two-finger tap to show the highlighter

- Two-finger pinch/unpinch to resize the highlighter

- Single-finger tap to hide the highlighter

This interface is actually quite simple to implement, but through trying various combinations, we developed a number of techniques for interpreting gestures.

Exploring the Multitouch API

To implement your own multitouch gestures, you need to:

- Arrange for touch messages to get routed to your code

- Understand the information passed to you

- Track and parse gestures from that information

Handling Events

If you're implementing your own custom controls, arranging for touch messages to get routed is easy; you just need to implement the relevant event handlers.

There's a set of four with identical signatures, declared in `UIResponder`:

```
- (void)touchesBegan:(NSSet *)touches withEvent:(UIEvent *)event;
- (void)touchesMoved:(NSSet *)touches withEvent:(UIEvent *)event;
- (void)touchesEnded:(NSSet *)touches withEvent:(UIEvent *)event;
- (void)touchesCancelled:(NSSet *)touches withEvent:(UIEvent *)event;
```

You also need to ensure your view's `userInteractionEnabled` is set, as well as `multipleTouchEnabled` if appropriate. UIViews and *most* subclasses default to `userInteractionEnabled`, but `multipleTouchEnabled` is usually disabled.

These four messages, plus the `UIEvent` class and the `UITouch` class, are the core of multitouch handling. The messages are pretty self-explanatory, but here's a quick note on the difference between `touchesEnded` and `touchesCancelled`: touches end when the user lifts a finger, but they're cancelled by the operating system itself, typically in response to asynchronous events such as incoming text messages or phone calls.

A `UITouch` object represents a single finger in contact with the screen. It's created when a finger touches the screen, destroyed when the finger leaves it and, notably, is attached for its entire lifetime to a single `UIView`. The view the finger initially came into contact with is the only view that receives the touch events. This is why, for example, you can touch the slot machine–style `UIPickerView` controls and drag your finger up and down across the entire screen height: the control captures the touch (see Figure 4-4).

The `UIEvent` represents an entire gesture or sequence, encompassing up to five fingers. It's created when the first finger comes into contact with the screen (and you receive your first `touchesBegan:withEvent:`), containing that finger's `UITouch`, and it continues to exist—with `UITouch`es being added or removed as additional fingers come and go—until all fingers are removed from the screen. Only then (after the final `touchesEnded:withEvent:` or `touchesCancelled:withEvent:` messages) is the `UIEvent` destroyed.

For each of the touchesX:withEvent:, the touches set will only contain the changed touches, whereas the UIEvent always contains the complete set of active (or, in the case of ended/cancelled events, recently active) touches. Note also that you don't receive a message specifically telling you an event has ended. Combining these facts, we can detect a UIEvent ending by checking ended/cancelled touches: if the size of the changed touches set is the same as the size of the event's touchesForView set, we know the last fingers are being lifted from the screen.

One of the issues that complicates multitouch handling is that the changed touches, the results of touchesForView, and the event's allTouches property are all unordered NSSets. If you need to track the movements of individual fingers, this can make life awkward, especially since UITouch objects can't be used as NSDictionary keys and, according to Apple, should not be retained. A UITouch object's address does remain constant throughout its life, so Apple recommends using these addresses as keys in a CFDictionary (not NSDictionary). It's worth remembering, though, that for many gestures, especially when tracking two fingers, this solution may be way more complex than you need.

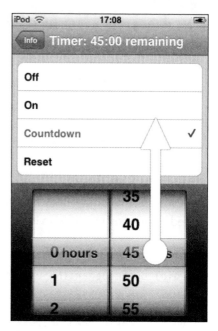

Figure 4-4. *Once a finger is placed on the scrolling view, it can be moved across the entire height of the screen to scroll the view, without disturbing the multiple-choice buttons above it.*

A good approach is often to simply ignore the ordering of the touches, or the specific UITouch-to-finger mapping and concentrate only on the relationship between the touches. For example, if you're handling a pinch/unpinch gesture, all you really care about is the distance between the two touches, and the order is irrelevant. When measuring rotation, we can use a simple pointer-compare to order the touches so that the rotation doesn't flip by 180 degrees if the provided ordering changes (this is shown in the sample code later in this chapter). It turns out, a multitouch control rarely needs to track an individual touch for its lifetime.

There's one final note on event handling before we start looking at specific gestures and detection techniques: if you're looking to modify existing controls, it can be tricky to get those events delivered to your code in the first place. If you subclass an existing view, you may find that you don't receive the events you're expecting, because many of the standard views are actually collections of views. When you create, for example, a UISlider, Cocoa actually creates several UIViews contained as subviews of the slider view.

Since we have no documented way of changing the class of these subviews, we can't subclass them and override their `touchesX:withEvent:` methods. The classes themselves may be undocumented, and developers are contractually obliged not to use undocumented code. Does this mean we're doomed?

Actually, no. Let's go back to your subclass that's not receiving events like it should. We find a useful method here:

```
- (UIView *)hitTest:(CGPoint)point withEvent:(UIEvent *)event;
```

If we override this, we can arrange for touches to be delivered to our class, not the subclass. We can then inspect the event, handle it if necessary, or, if we want to retain standard behavior, simply forward the messages to their original destinations. Don't make assumptions about the internal structure of these compound views, as Apple has left them undocumented and could change them in future updates. However, by using our superclass's `hitTest:withEvent:`, we can ensure the events get delivered to the right place without needing to know the structure ourselves.

Recognizing Gestures

How do we interpret gestures? It's an interesting question, because it's one of the closest things we do to mind reading. Gestures are not explicit like mouse actions such as clicking a button or dragging icon A over widget B. As a result, we're going to rely heavily on heuristics and approximations to get us through, but we'll also think about the user's mental model.

Let's look at some iPhone gestures used by the built-in applications:

- *Tap*: Used throughout to activate buttons
- *Double-tap*: Not that common but used in Safari to zoom in and out of paragraphs
- *Finger scroll*: Used often to scroll through web pages, tables, contact lists, and so on
- *Swipe*: Used to switch pages in the Weather application, move from picture to picture in the Photos application, and erase items from lists (e.g., to delete spam e-mail)
- *Pinch/unpinch*: Used to zoom in and out of photos, documents, and web pages
- *Two-finger scroll*: Rarely used, most notably used for scrolling inside frames of Safari

Many of these are quite simple to detect individually: it's easy to find out how many fingers are on the screen; a double-tap is tracked for you (`UITouch` events have a `tapCount` property); Apple includes an example of one version of swipe detection in the SDK documentation; pinch/unpinch can be tracked with a little Pythagorean math. It's the subtleties that are important.

Apple's built-in controls already understand these gestures where appropriate, but sometimes, you'll find yourself needing to reimplement them. If you do, you need to be careful with the details: for example, list scrolling can be flicked to scroll rapidly then slow down to a gentle stop, but if you keep your finger on the screen, the list's position tracks the finger precisely. Other examples are the way the list bounces at the top and bottom edges, emphasizing the physicality of the interface and the way you can drag lists beyond their ends (this is a definite no-no on the Mac, but on the iPhone, Apple seems to have judged that maintaining the link between finger and page is more important than the aesthetics, as demonstrated in Mobile Safari when it has to checkerboard a web page to keep up).

Where gesture recognition gets particularly tricky is if a single view can accept multiple gestures and you need to decide, often before a gesture is complete, how to interpret the user's input. For example, a pinch gesture involves two fingers on the screen, moving linearly. A two-finger scroll gesture also involves two fingers on the screen, moving linearly, and (since people usually don't have robot-perfect motor control) those fingers might get a little bit closer or further away from each other in the process.

How do we resolve these ambiguities and make an interface feel responsive?

It's important to be generous in your interpretation, but take context into account here. Stage Hand is stricter with its detection, because of the principle of least embarrassment and because it's used in a fairly controlled environment (on stage, in a lecture theatre, in a classroom or boardroom). If you're writing an application someone will be using on a bus or while walking around, and where the cost of a misread command is low, you should be far more generous.

Many touch interfaces rely on physical metaphors: your finger interacts with controls as though they were physical machinery with weight, momentum, and inertia. While your finger is touching the controls, they try and match its position exactly, but when released, they may spin, spring, or glide to a halt as appropriate.

This applies to other gestures too: one mistake we made in an early Stage Hand build was to fail to take inertia into account for dragging the slates around. The basic behavior we had was that a slate would sit centered on the screen until dragged by the user. It would follow your finger exactly, and when you released it, it would either snap back to its original location, or if it was more than halfway off the screen, the next slate (which would have been pulled into view by this point and should be taking up the majority of the screen) would snap in instead (see Figures 4-5 and 4-6). You're probably familiar with this behavior from the Home screen if you have more than one page of applications.

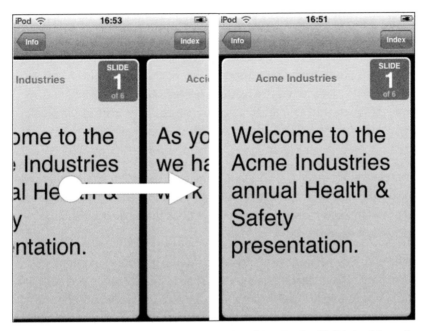

Figure 4-5. *If the slate is held still and released at this point, it will slide back into place, as the next slide's threshold has not yet been crossed.*

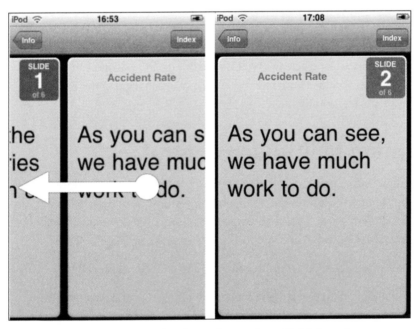

Figure 4-6. *Here, the second slide's slate is the one that will slide into place.*

The problem was that if you flicked the slate a little bit, it would not move far enough to change slides. This seems reasonable until you consider the weight and inertia of the slate. When flicked, people subconsciously expect the slate to keep going until friction slows it to a stop. So, it's not the slate's position when the finger is released that we should be checking, it's the position where the slate comes to a halt. But, we also want to be responsive, so we don't want to wait for the slate to animate to a stop before deciding. Instead, we predict based on the slate's speed where it will end up and take the decision immediately. This helped enormously.

This example illustrates what we mean by considering the user's mental model: trying to figure out what, even subconsciously, people expect the interface to do when poked. Because the finger-on-screen interaction is more direct and visceral than the indirect poking of a mouse, users have different expectations.

Sometimes, you need to decide what the user's trying to do, and then rule out other options (at least, until enough fingers have cleared the screen for you to reset). For example, as well as swiping slates from side to side to change slides, you can also scroll your notes up and down. There's no technical reason we couldn't allow your finger to move in two dimensions, simultaneously scrolling notes and dragging the slate to one side. But there's also no earthly reason why a user would want to do that, and it would drive users crazy if their notes were slipping and sliding from side to side while they tried to scroll them.

As a result, we track user movement, quickly decide whether the intention is to scroll or drag, and lock the behavior accordingly. The result feels much more solid and reassuring. It's like the difference between walking across a typical road bridge and walking across one of those wood-plank-and-rope bridges from Indiana Jones movies, shaking under every footstep. We want our users to feel confident!

Implementing Multitouch Controls

The sample application shows how to interpret a variety of multitouch actions, decide between ambiguous ones, handle multiple simultaneous gestures, and correctly handle gestures happening in different views. It presents two pinball-bumper–style graphical gadgets sitting on a backdrop (see Figure 4-7) and supports the following commands:

- Swipe the backdrop up and down with your finger to change backdrop image.

- Use a single finger to drag the gadgets around, as if they attached directly to your finger.

- Flick the gadgets around with simple physical simulation.

- Pinch/unpinch to shrink or stretch the gadgets.

- Double-tap a gadget to reset it to its default size.

- Rotate the gadget with two fingers.

- Two-finger drag to lock the movement of the gadget to one axis.

Additionally, one of the gadgets offers free-form stretch and rotate, while the other is mode-locked, where you can either rotate or stretch it but not both at the same time; we use the view's Tag property to set this (see Figure 4-8). If you're building along at home, check Figure 4-9 to see how the Interface Builder document is set out.

Figure 4-7. *The bumper in the top-left corner can be rotated, moved, and scaled simultaneously. The bumper in the bottom-right corner locks into one mode at a time.*

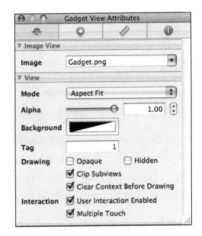

Figure 4-8. *We set the Tag of the bottom-right bumper to 1, to indicate this is the one that exhibits mode-locked behavior (the other is left at the default, 0).*

Figure 4-9. *Note that the two GadgetView objects are siblings of the BackgroundView, not children of it. This ensures they're not disrupted when the background image changes.*

Handling Touches

Let's get into the implementation. In Interface Builder, I placed two UIImageView instances (Figure 4-10) and changed their class to my custom GadgetView subclass (Figure 4-11).

There's no custom drawing code or anything like that: the only changes are to support the multitouch gestures. Most of the work is in handling the four multitouch events we described earlier, for touches beginning, moving, ending, and cancelling. We'll also sometimes be doing some animation, triggered by a timer, which we'll start and stop as necessary.

When you place a finger on one of the gadgets, touchesBegan:withEvent: is called:

```
- (void)touchesBegan:(NSSet *)touches withEvent:(UIEvent *)event
{
    NSSet* allTouches = [event touchesForView:self];
```

The first thing we do is get all the touches that match this view. The touches parameter only contains the touches that have just hit the screen, and we need to account for users whose fingers hit the screen one slightly later than the other, as well as the ninjas whose fingers are exactly in sync.

```
    if ([allTouches count]>2)
    {
    }
```

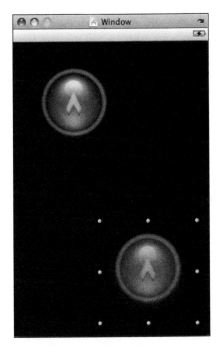

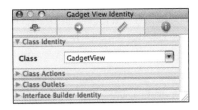

Figure 4-10. *The two bumpers were created by simply dragging the images from the Interface Builder Media library into the document window.*

Figure 4-11. *The bumpers' class was then changed from UIImageView to GadgetView using the Identity Inspector.*

We don't process any gestures for three or more fingers, so we filter them out early. There's no reason you couldn't add your own processing here, at least, no technical reason. Whether it's advisable or not from a user-experience perspective is another question.

So, this is how we handle a two-finger gesture:

```
else if ([allTouches count]==2)
{
    NSArray* twoTouches = [allTouches allObjects];
```

NSSet is unordered; we get the touches into an array, which doesn't sort it for us, but at least we can address each touch individually even if we don't know which is which yet.

```
    lastPinch = [self calculatePinch:twoTouches];
    lastRotation = [self calculateAngle:twoTouches];
    lastCenter = [self averageTouchPoint:twoTouches];
```

We'll look at the implementations of these methods later; the important thing is that we're initializing three tracking variables: a floating-point number to track the distance between the fingers, another to track their relative angle, and a CGPoint to track the midpoint between them.

We'll also need to track whether we've locked the interface into a particular gesture. When performing two-finger drags we'll set the modeLock to either lockToXAxis or lockToYAxis, and when interacting with the mode-locked gadget, we may set it to lockToRotation or lockToScale. Until then, it defaults to lockNotYetChosen:

```
if (modeLock==lockNotYetChosen)
{
    axisLockedDrag = YES;
    originalCenter = lastCenter;
}
```

Until we get some sign that the user's performing a more specialized gesture, we can only assume that two fingers on the screen could mean a two-finger drag. We'll initialize for this, but note the test we do first, as it's a subtle but important one: modeLock is set once a gesture has been detected and locked into place—and only cleared when all fingers involved in the gesture have left the screen. So, this test means you can temporarily lift and replace one finger and not reset the gesture. It's particularly important, because it's very easy for a finger to brush over the edge of the display when dragging or flicking, and the iPhone will register this as the touch ending. We need to account for this situation; otherwise, we'll get really unpleasant behavior near the edge of the screen.

That's all for two-finger gestures, we just need to handle the simple one-finger case now:

```
    }
    else
    {
        UITouch* touch = [allTouches anyObject];
```

We just grab the only touch there is, and it turns out we don't actually need to initialize any state for single-finger drag operations. We just check for double-taps, which reset the gadget to its default size (stored earlier, in awakeFromNib, not shown here):

```
        if (touch.tapCount==2)
        {
            CGPoint ctr = self.center;
            CGRect bounds = self.bounds;
            bounds.size.width = defaultSize;
            bounds.size.height = defaultSize;
            self.bounds = bounds;
            self.center = ctr;
        }
    }
}
```

Deciding What Movement Means

What about when a finger moves? Just like before, we get all the view-specific touches and see how many there are to determine our response. The handling for each case is more complex, though:

```
- (void)touchesMoved:(NSSet *)touches withEvent:(UIEvent *)event
{
    NSSet* allTouches = [event touchesForView:self];
    if ([allTouches count]==1)
    {
        if (modeLock>lockNotYetChosen) return;
```

This test is another part of dealing with finger-ups at the edge of the screen: if one of the fingers in a two-finger gesture temporarily loses contact with the screen, we don't want to revert to single-finger drag behavior, so we check and lock the behavior accordingly.

But, if we're just doing a regular drag, the code is pretty simple. Touches remember their last position as well as their current one, so we just get both these positions and subtract one from the other to get the finger's relative movement:

```
        UITouch* anyTouch = [touches anyObject];
        lastMove = anyTouch.timestamp;
        CGPoint now  = [anyTouch locationInView: self.superview];
        CGPoint then = [anyTouch previousLocationInView: self.superview];
        dragDelta = CGPointDelta(now, then);
        self.center = CGPointApplyDelta(self.center, dragDelta);
        [self stopTimer];
}
```

Notice that we don't get the touch's location in the current view! This is because we're moving the view itself, which means the coordinate system would be continuously changing and drive us nuts. I mean, since we're dragging the view along with the finger, if we're doing our job right, the finger shouldn't move relative to the view at all! So, we find the finger's position in the superview instead, which gives us a fixed frame of reference.

This technique applies to any animated view, although bear in mind that for many animated views, you'll want to take the simpler approach of just disabling interaction with a view while it's animating.

We also record when the touch occurred, as we'll need this later. Once we've done all this, all we need to do is update the graphics to match. We also stop any background animation that's occurring on this view (of which more later in the section "Applying Weight and Inertia") since the user's finger movement takes priority.

Now for the fun, two-finger gestures. We start the usual way, by collecting together all the data we may need:

```
else if ([allTouches count]==2)
{
    NSArray* twoTouches = [allTouches allObjects];
    float pinch = [self calculatePinch:twoTouches];
    float rotation = [self calculateAngle:twoTouches];
    CGPoint touchCenter = [self averageTouchPoint:twoTouches];
    CGSize delta = CGPointDelta(touchCenter, lastCenter);
    CGPoint gadgetCenter = self.center;
```

This block is for the first few frames of the gesture, before we know what kind it is. It's going to keep an eye out for clues as to the user's intent.

```
    if (axisLockedDrag && (modeLock==lockNotYetChosen))
    {
        if (fabsf(pinch-lastPinch)>PINCH_THRESHOLD)
        {
            axisLockedDrag = NO;
            if (modal)
                modeLock = lockToScale;
        }
        else if (fabsf(rotation-lastRotation)>ROTATION_THRESHOLD)
        {
            axisLockedDrag = NO;
            if (modal)
                modeLock = lockToRotation;
        }
```

If the fingers get closer together or further apart, or rotate relative to each other, we don't have a two-fingered drag (which assumes constant relative position between the fingers). Of course, a little wobble is to be expected, so the relative motion has to exceed a threshold. If this happens, we disable dragging. One of the gadgets, if you remember, allows free-form editing (you can scale and rotate at the same time), but the other locks itself to whichever happens first.

If the fingers are moving together, we're still doing a drag operation, but again, we check against thresholds. Once we're sure the user has a particular axis in mind, we lock to it:

```
        else
        {
            CGSize dragDistance = CGPointDelta(touchCenter, ➥
                                     originalCenter);
            if (fabsf(dragDistance.width)>AXIS_LOCK_THRESHOLD)
            {
                modeLock = lockToXAxis;
```

```
        }
        else if (fabsf(dragDistance.height)>AXIS_LOCK_THRESHOLD)
        {
            modeLock = lockToYAxis;
        }
    }
}
```

Applying the Movement

Now we know what's going on! It's just a matter of putting the appropriate gesture into effect, depending on which we've detected:

```
if (axisLockedDrag)
{
    switch(modeLock)
    {
    case lockToXAxis:
        delta.height = 0;
        gadgetCenter.y = Interpolate(gadgetCenter.y,
                                        originalCenter.y, 0.1f);
        break;
    case lockToYAxis:
        delta.width = 0;
        gadgetCenter.x = Interpolate(gadgetCenter.x,
                                        originalCenter.x, 0.1f);
        break;
    }
    self.center = CGPointApplyDelta(gadgetCenter, delta);
}
```

The axis-locked drag is the same as the single-finger drag from before, except we cancel out the motion on whichever axis we haven't locked to. Also, because we don't know which axis to lock to until the user starts moving that way, we allowed some free-form dragging at first. We need to cancel this out, but simply snapping the gadget into place is jarring. We linearly interpolate it back instead.

Another approach is to simply refuse to move the gadget at all until an axis has been locked in. This method is less responsive, so the user will perceive a need to pull harder on the gadget before it will start moving, but the interface can feel more stable or even slightly magnetic, like a MagSafe power adapter.

Here, we start to handle the various transformations that can be applied to the gadget:

```
else
{
    if (modeLock!=lockToScale)
```

```
        {
            CGAffineTransform transform = self.transform;
            transform = CGAffineTransformRotate(transform,
                                            rotation-lastRotation);
            self.transform = transform;
        }
```

Note that rather than testing for the current lock, we apply any transformations that aren't locked out, since in free-form mode several transformations may be applied at once.

```
        if (modeLock!=lockToRotation)
        {
            float scale = pinch/lastPinch;
            CGRect bounds = self.bounds;
            bounds.size.width *= scale;
            bounds.size.height *= scale;
            self.bounds = bounds;
        }
```

There's one more thing to take care of in free-form mode: when you're rotating and scaling a gadget, you may not move both fingers evenly. In fact, you can slip-and-slide all over the place, and because we're only tracking relative motion, the control will still work. But it'll look weird. If we update the position of the gadget to the midpoint of the fingers, it'll maintain the illusion that we're manipulating a physical object:

```
        if (modeLock==lockNotYetChosen)
            self.center = CGPointApplyDelta(self.center, delta);
```

Finally, update some of the tracking state, so that we make the comparisons against the right values next time round. And we're done!

```
            lastPinch = pinch;
            lastRotation = rotation;
        }
        lastCenter = touchCenter;
    }
}
```

Applying Weight and Inertia

We've now implemented all the gestures, but there's something missing. I mentioned earlier the importance of weight and inertia, and it's time to put it into practice. If we flick an object, it should keep going.

All the work for this is done at the moment the finger is lifted, starting off as usual by acquiring the current touches. Remember, at this stage, the touches being removed from the screen are still included. We put into practice the standard check for all fingers being removed: if any fingers still touch the view, it's considered as being gripped, and we leave it alone.

```
- (void)touchesEnded:(NSSet *)touches withEvent:(UIEvent *)event
{
    NSSet* allTouches = [event touchesForView:self];
    if ([touches count]==[allTouches count])
    {
        modeLock = lockNotYetChosen;
```

The gadget's been released. The obvious thing is to cancel the lock at this point, but before we can implement inertia, we must check when was the gadget last moved:

```
        if ((event.timestamp - lastMove) > MOVEMENT_PAUSE_THRESHOLD)
            return;
```

If the gadget has been stationary for a while, it shouldn't start moving just because we let go of it. It's also common for your finger to roll or shift position slightly as you lift it off the screen, so we also ignore small movements:

```
        if ((fabsf(dragDelta.width)>INERTIA_THRESHOLD) ||
         (fabsf(dragDelta.height)>INERTIA_THRESHOLD))
        {
            [self startTimer];
        }
    }
}
```

The gadget needs to move without the user's interaction, so we run a timer at 60 frames per second (the iPhone's screen refresh rate) to update it. I assume you know how to use NSTimer by now, so we'll just look at what happens when the timer fires:

```
- (void) timerTick: (NSTimer*)timer
{
    dragDelta = CGSizeScale(dragDelta, INERTIAL_DAMPING);
```

When we calculated the user's finger movement earlier, we stashed it in a member, rather than local, variable so that we'd have access to it later. This way, the gadget keeps moving at the speed the user last moved it. Each time the timer fires, we multiply this vector by a fraction a little under one to simulate the friction of the surface below the gadget.

This simple simulation would never quite reach zero, so we only process it if it's above a certain level:

```
        if ((fabsf(dragDelta.width)>DELTA_ZERO_THRESHOLD) ||
             (fabsf(dragDelta.height)>DELTA_ZERO_THRESHOLD))
        {
            CGPoint ctr = CGPointApplyDelta(self.center, dragDelta);
```

We calculate the new position of the gadget, but we don't apply it right away. We need to test it first; otherwise, it's really easy to just flick a gadget into outer space. Just like a scrolling list, if our gadget does hit the edge, it'll bounce off. Collision detection's a little outside the scope of this chapter, but you can look it up in the sample code if you need to.

```
        CGSize halfSize = CGSizeMake(self.bounds.size.width/2,
                        self.bounds.size.height/2);
        [self check:&ctr delta:&dragDelta halfSize:halfSize
                    forBouncingAgainst:self.superview.bounds.size];

        self.center = ctr;
    }
    else
    {
```

If the inertial movement drops below a noticeable level, we just kill it off outright—saving the user's batteries, apart from anything else. You'll remember from earlier that we also stop it if you place a finger on the gadget, allowing you to trap it in flight.

```
        [self stopTimer];
    }
}
```

Tying Up Loose Ends

We've now looked at the processing for every part of the gesture, and by now, you should be ready to implement your own. We'll just close this section by noting that we also clear the gesture lock if a touch is cancelled by the operating system and, as promised earlier, by having a quick look at how we calculate the values used to track the fingers.

The calculateAngle method is the most complex of the set, so we'll pick it apart first:

```
- (float) calculateAngle:(NSArray*)twoTouches
{
    NSParameterAssert([twoTouches count]==2);
    UITouch* firstTouch = [twoTouches objectAtIndex:0];
    UITouch* secondTouch = [twoTouches objectAtIndex:1];
```

After a quick safety check, we simply pull out the two touches and arbitrarily decide that one of them is the first. We poured them into an array so we could use objectAtIndex, but the array itself can be in arbitrary order. This is the only method that actually cares what order the touches come in, so we simply compare the pointers and swap them if necessary:

```
    if (firstTouch>secondTouch)
    {
        UITouch* temp = firstTouch;
        firstTouch = secondTouch;
```

```
        secondTouch = temp;
    }
```

The same UITouch object representing each touch remains the same throughout the event, so if we sort them like this, we know we're preserving an order—any order. The order doesn't matter as long as it doesn't change midgesture. Having done this, we can extract the coordinates, again in the parent view's reference frame so that the rotation of the gadget itself doesn't throw our calculations off:

```
    CGPoint first  = [firstTouch locationInView:self.superview];
    CGPoint second = [secondTouch locationInView:self.superview];
    CGSize  delta  = CGPointDelta(first, second);
    return atan2f(delta.height, delta.width);
}
```

Finally, an atan2f from the C standard library gets us the angle described by the line between the points.

The other calculations are equally straightforward and follow the same patterns, so there's no need to rehash them. The calculatePinch method just gets the length of the line, instead of the angle. averageTouchPoint is even simpler, calculating the mean of the array of coordinates passed to it.

Summary

It's important to remember that the sample code in this chapter is an example of the techniques, not of good application design! It's rarely a good idea to overload a single gadget with this many behaviors. Read Apple's *iPhone Human Interface Guidelines* (log in to http://developer.apple.com/iphone/ to find this document). Don't think of them just as some rules to stick to, but consider the decisions behind them. It's the spirit, not the letter, that is important. iPhone applications are not only different from desktop applications but also have a wider range of variation in their own right, because they're not all used seated at a flat, controlled surface. Some, like Stage Hand, may be used in a relatively controlled environment but by a user under stress. Others may be purely for relaxation but could be used while being jolted around on public transit with a full bag of groceries in the other hand.

It's far better to design something cleanly in the first place than to rely on fancy tricks to get you out of trouble. Sometimes, though, you are painted into a corner or have no choice, and it's better to have these tools at your disposal, especially if they're for power or optional, extra features rather than the application's core task.

If you have to use advanced gestures, try to keep the processing as simple as possible. It's easy to overthink it! You don't need to write a handwriting-recognition system. Instead, focus as much as possible on the user's intent and do the minimum to detect that intent and differentiate it from others. Definitely don't try to be too restrictive and require too much accuracy from the user.

Users sometimes do the strangest things, but if you can put yourself in their places (or if you have the opportunity to ask them) and find out why, the results can be very illuminating. What's best is when you can put some of that understanding into the application. When you consider why your users want to do something, you can be a step ahead of them and present sensible choices, or enable only the appropriate gestures, which typically increases the reliability of detection. You don't have to be psychic, but by putting yourself in the user's head, you can make the iPhone seem to be!

Benjamin Jackson

Company: Brainjuice

Location: Rio de Janeiro, Brazil

Former life as a developer: Languages, roughly in order: BASIC, Pascal, C/C++, Objective Caml, Java, Flash/ActionScript, and Ruby/Rails. Development environments: Eclipse, Visual Studio, and Flash. Programs used: TextMate, Creative Suite, and Terminal.app.

Life as an iPhone developer: Created the Arcade Hockey game and Town Hall reference application

iPhone development toolset: TextMate for editing, Xcode for building, and instruments and clang static analyzer for detecting leaks

What's in this chapter: This chapter covers setup with cocos2d, responding to touch events, and collision detection

Key technologies

- *cocos2d*

- *Chipmunk*

- *OpenGL ES*

Chapter 5

Physics, Sprites, and Animation with the cocos2d-iPhone Framework

o you want to write an iPhone game with realistic physics. This chapter will get you started.

Here's our story: Brainjuice was founded in 2007 by myself and Ivan Neto as the product division of our agency, INCOMUM Design & Concept. That December, we launched Blogo, a weblog editor and our first Mac desktop application.

After the first version of the iPhone SDK was released, developing native applications for the phone was a natural next step. We were already well versed in Objective-C, and the freedom to mix C with Objective C when needed (as well as take advantage of all of the existing C libraries out there) is a big help. The touch screen interaction is unmatched on any other device, and when you add in the three-way accelerometer, the possibilities for different interfaces and controls are endless. Gestural controls mixed with realistic physics can create a scarily immersive game experience.

We made the choice to hold off an iPhone version of Blogo, which we knew would be a complex problem both in terms of the interface design and the number of different services we would have to integrate with. Instead, we wanted to cut our teeth on a game, which would help us get a feel for developing in a multitouch environment. Our experience at the time included 2D scrollers with Flash, 3D augmented reality simulation, and virtual world development on multiple platforms, but we still hadn't done anything with two-player competition.

Air hockey is an arcade classic with mass appeal. We all played it as kids, and none of the existing hockey games in the App Store was up to our standards in terms of playability and realism. Factors like the goal size, the friction on the table, and the size of the mallet have a huge impact on the game play experience, and none of the applications we tested had the right combination. Most were free.

In this chapter, I'm going to explain some of the concepts behind 2D game programming. To illustrate, I'll discuss the process we went through developing Arcade Hockey, including the challenges we faced and how we got around them. I'll explain lighting, collision detection, physics simulation, and how to get a game from prototype to playable.

Getting Started with Game Programming

Programming a game is fundamentally different from other types of applications, on or off the iPhone. Rather than designing a series of screens for the user to navigate, the developer has to design a virtual world for the user to interact with. This space can be either 2D or 3D, and the programmer's task is twofold:

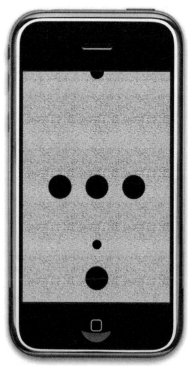

- Design the space, and define its constraints. These can be the outer boundaries of the space as well as the internal ones (for example, in a maze game), but constraints are not limited to position. Friction, elasticity, gravity, and even lighting can all play a role in defining the user's experience in the virtual space.

- Define rules for how objects interact with the space. This generally comes down to defining what happens when objects collide with each other. For example, a user might score when the ball hits the back wall of a goal or die when the player comes into contact with a lethal object.

In this chapter, we're going to look at the unique challenges presented by developing a 2D game on the iPhone. To do this, we're going to use OpenGL ES with help from the cocos2d and Chipmunk libraries to create a simple miniature golf game (as shown in Figure 5-1).

Figure 5-1. *The miniature golf game layer*

Introducing OpenGL ES

OpenGL ES is the most widely used graphics programming API today. Implemented as a standard C library, it forms the basis for countless 2D and 3D games and development frameworks. On systems with processors optimized for OpenGL, the heavy computation is passed off to the graphics card, resulting in much better performance and less load on the CPU.

OpenGL ES is a subset of the OpenGL API optimized for embedded devices like mobile phones, PDAs, and handheld video game consoles like the Nintendo DS and PlayStation Portable (PSP).

Introducing cocos2d and Chipmunk

OpenGL is powerful, but hand-coding the same calls to its cryptic method names repeatedly will have any sane person leaving head-shaped holes in the wall before the first proto-type is out the door. Enter cocos2d and Chipmunk—two libraries that take the raw power of OpenGL and add a layer of abstraction to package it up into the language of 2D game programming.

In 3D game programming, rendering is based on sets of 3D coordinates mapped to **textures**. In 2D game programming, we base the display on 2D coordinates mapped to **sprites** (flat graphics moved in the two-dimensional space). Sprites are organized in **layers** to allow for simple stacking on the Z axis. cocos2d works with 2D **vectors** (sets of coordi-nates) to define both points on the stage and transformations on those points (for example, "move vector X by vector Y"). We use vectors to define our sprites' positions.

To simulate real-world physics, we piggyback Chipmunk on top of cocos2d. Chipmunk adds functions that let us define the physical properties of our sprites (friction and elasticity), as well as how they interact with each other when they collide.

Developing Arcade Hockey

In this section, we're going to discuss the process behind creating Arcade Hockey (shown in Figure 5-2). During the development, we learned a lot about what it takes to make a game not only playable but fun, and we ran into more than a few stops that required some cre-ative workarounds.

Figure 5-2. *The main menu of Arcade Hockey*

We began by studying the other hockey games in the App Store, and a couple of paid competitors gave us an early scare. We downloaded and tested all of the other applications to see what the competition had come up with already.

We noticed that the goals were all so small that scoring was nearly impossible if the opponent kept the mallet front and center. We also noticed that the artificial intelligence's strategies invariably revolved around keeping the mallet front and center—no coincidence there. After we had an idea of what needed improvement, we moved on to sketching out the screens and basic options.

For the brand and interface design, we chose a creative direction based around the old-school 80s video-game graphics, which were the norm in the arcades where air hockey became popular. All graphics were done as vector illustrations in Adobe Illustrator, which allowed us the liberty of scaling them for multiple uses without losing quality. For example, Arcade Hockey offers multiple mallet and puck sizes (see Figure 5-3). We mocked up a prototype with a bare table and placeholder graphics to begin working on the physics while the sprites were being designed.

Figure 5-3. *The Options screen*

Once the mallets felt right, we began sketching out the basic game play with the puck. We quickly learned that collision detection was far from plug and play, and we began adjusting the code to avoid bugs like the multiple collisions that would pile up when the puck was moving slowly and barely touched a mallet or when pressing the puck up against a wall. To keep these kinds of bugs from hitting your own application, make sure to set a flag when the first collision is detected and set a timer to clear it after a fraction of a second.

From there, we moved on to the artificial intelligence (AI). This turned out to be a huge part of the work, and we went through several different iterations with varying strategies until we got the AI smart enough to present a real challenge.

For example, how do you know whether or not the enemy is on the attack or on the defensive? One way is to look at which half of the table the puck is in. But this doesn't take into account cases where the puck is in the enemy's area but moving toward the goal. In the end, the best strategies combine knowledge of the puck's position relative to the mallet and its velocity.

Finally, we fleshed out the Options screen, as shown in Figure 5-3, by adding different puck and mallet sizes and added the finished art to the program, as well as some final touches like the simulated game noises in the first screen, and the flaming puck that appears in a sudden-death match point. We also did some fine-tuning on the game play, adjusting the maximum speed of the puck to keep the game from getting out of hand.

Next, I'll go into some more detail on a couple of the specific challenges we faced and show you how we solved them.

Tracking the User's Finger

Once we were ready to start coding, we started by getting the mallets working and tracking the user's finger. This proved to be a bit of a difficult problem because of an important usability issue: the user needed to be able to see the mallet while controlling it at the same time. We found, after playing around for a bit, that the best position for the mallet is slightly in front of the finger.

However, this meant that there was no way for the user to protect the goal when pulling back to defend. In the end, we had to adjust the puck's position relative to the user's finger so that, when at the back of the screen, it would be directly underneath (as shown in Figure 5-4), and gradually move forward as the mallet approached the center line.

Figure 5-4. *Finger tracking while guarding the goal (shaded circles show finger positions)*

Listing 5-1 shows a code snippet for compensating for the user's finger.

Listing 5-1. *Compensating for the User's Finger*

```
int finger_padding = 30;
int fat_fingers_offset = 40; // we add a little bit more to the offset

// ... snip
```

First, we set some values, which we'll use later to offset the puck in relation to the user's finger:.

```
- (void)touchesBegan:(NSSet *)touches withEvent:(UIEvent *)event {
  for (UITouch *myTouch in touches) {

    CGPoint location = [myTouch locationInView: [myTouch view]];

    // translate location coordinates into game coordinates
    location = [[Director sharedDirector] convertCoordinate: location];
```

Next, we get the touch location relative to the view. We need to convert this coordinate relative to the current layer using the convertCoordinate: method in the Director object:

```
    if (CGRectContainsPoint(player1AreaRect, location)) {

      // don't let the mallet pass into the other player's area
      if (CGRectContainsPoint(notPlayer1AreaRect, location)) {
        break;
      }
```

We block the mallet from hitting any point outside its area and break out if we find it inside notPlayer1AreaRect (defined earlier in the code):

```
      // calculate the padding based on the location
      cpFloat padding = finger_padding * ((120 - location.y) / 100);
      location.y -= padding;
      location.y += fat_fingers_offset;
```

This is the trick: we cut the mallet's y value down by a factor that approaches zero as the finger reaches the bottom edge of the table and increases linearly to a maximum of 80 pixels ahead of the finger at midfield:

```
      // lock the location to the center line
      if (location.y > 240) location.y = 240;
      cpVect p1position = cpv(location.x, location.y);
      cpMouseMove(player1Mouse, p1position);
  }
```

Since we're pushing the puck ahead, we want to lock it to the middle of the table as well:

```
else if (CGRectContainsPoint(player2AreaRect, location) &&
        !singlePlayerMode) {

    // don't let the mallet pass into the other player's area
    if (CGRectContainsPoint(notPlayer2AreaRect, location)) {
      break;
    }

    // calculate the padding based on the location
    cpFloat padding = finger_padding * (( location.y - 360 ) / 100);
    location.y += padding;
    location.y -= fat_fingers_offset;

    // lock the location to the center line
    if (location.y < 240) location.y = 240;
    cpVect p2position = cpv(location.x, location.y);
    cpMouseMove(player2Mouse, p2position);
  }
 }
}
```

We then do the same for the second player, unless we're in single-player mode.

Detecting Collisions

Collision detection is at the heart of any game that attempts to simulate reality. In the real world, objects bounce off each other with a specific direction and velocity depending on the inertia of the two bodies and their speed and direction.

In most games, we're not only interested in the objects colliding realistically but also in triggering events based on those collisions (for example, a rocket colliding with its target). To handle these special cases, we attach callbacks (stored as C function pointers) to event types (stored as C constants).

We were surprised by the difficulty of getting collision detection working and playable. In particular, getting the mallets to continue on their trajectories after the user released them was impossible until we noticed on the mailing list that Chipmunk, the physics library we were using, had been updated with the new cpMouse functions. These functions simulate a mouse pointer in 2D space, allowing you to programmatically move an object while retaining its momentum after being released and were essential for the mallet to move realistically under the user's finger.

Sometimes, the puck would go so fast that it passed the wall without generating any collision event at all. After banging our heads against the wall trying to figure out how to prevent this, we ended up treating it as the virtual equivalent of the puck flying off the table with a careless hit.

We also noticed that the corners of the table would end up trapping the puck and made it impossible to continue play without pushing it off the table (as shown in Figure 5-5).

Figure 5-5. *Avoid getting the puck trapped in the corner.*

Giving the puck a small impulse toward the center when it hit the corners was enough to fix the problem, as shown in Listing 5-2.

Listing 5-2. *Keeping the Puck Out of the Corner*

```
enum {
  kColl_Puck,
  kColl_Goal,
  kColl_Horizontal,
  kColl_Player1,
  kColl_Player2,
  kColl_Pusher
};

// snip... initialize other variables
```

The first step to detecting collisions is initializing the types of collisions in an enum. We'll use these later to associate collision areas with C functions.

```
@implementation GameLayer

- (id) init {

  // snip... sets up the board, players etc.

  // add the puck
  puck = [self addSpriteNamed:@"puck.png"
    x:160 y:240 type:kColl_Puck];
```

First, we add the puck using addSpriteNamed:, the convenience function that follows. It takes an image, its dimensions, and a collision type (in this case kColl_Puck).

```
  // and a shape to the corner which will push it out
  cpShape *pusher = cpCircleShapeNew(staticBody, radius,
    cpv(distance, distance));
  pusher -> collision_type = kColl_Pusher;
  cpSpaceAddStaticShape(space, pusher);
```

Next, we add a collision shape to the corner and give it the kColl_pusher type we defined earlier.

```
  // finally add a collision pair function which
  // will be called when they collide
  cpSpaceAddCollisionPairFunc(space, kColl_Puck, kColl_Pusher,
    &puckHitPusher, puck);
}
```

Finally, we associate the collision area with the C function. To do this, we call cpSpaceAddCollisionPairFunc and send it our space (defined in the GameLayer's init function with cpSpaceNew()), the collision types that we want to associate, a reference to the C function, and a reference to the puck (this is not used, and could be NULL, but it doesn't hurt in case we need it later).

```
- (void)pushPuckFromCorner{
  cpBodyApplyImpulse(puck, cpvsub(cpv(160,240), puck->p), cpvzero);
}
```

Here's the key to the fix: we define a function to push the puck from the corner that will be called later from our C function callback. This just applies an impulse on the puck toward the center of the table.

```
- (cpBody *) addSpriteNamed: (NSString *)name x: (float)x
    y:(float)y type:(unsigned int)type {

  // add a new sprite and center it
  Sprite *sprite = [Sprite spriteFromFile:name];
  [self add: sprite z:2];
  sprite.position = cpv(x,y);
```

Now, we get to the convenience function we used earlier. We start by creating and position-ing a new sprite.

```
  // set up the vertexes based on the image dimensions
  UIImage *image = [UIImage imageNamed:name];
  int num_vertexes = 4;

  cpVect verts[] = {
    cpv([image size].width/2 * -1, [image size].height/2 * -1),
    cpv([image size].width/2 * -1, [image size].height/2),
    cpv([image size].width/2, [image size].height/2),
    cpv([image size].width/2, [image size].height/2 * -1)
  };
```

We then set up its collision vertices based on the image dimensions, centering the box on the center of the sprite.

```
  // every object needs a body
  cpBody *body = cpBodyNew(1.0, cpMomentForPoly(1.0, num_vertexes,
    verts, cpvzero));
  body->p = cpv(x, y);
  cpSpaceAddBody(space, body);

  // as well as a shape to represent its collision box
  cpShape* shape = cpCircleShapeNew(body, [image size].width / 2,
    cpvzero);
  shape->e = 0.5;  // elasticity
  shape->u = 0.5;  // friction
  shape->data = sprite;
  shape -> collision_type = type;
```

We give the object a body and a shape with the vertices we defined and set its elasticity, fric-tion, sprite data, and collision type.

```
  cpSpaceAddShape(space, shape);
  return body;
}
```

Finally, we add the shape to the space so it will show up on our main layer.

@end

```
int puckHitPusher(cpShape *a, cpShape *b, cpContact *contacts,
    int numContacts, cpFloat normal_coef, void *data) {
    [(GameLayer *) mainLayer pushPuckFromCorner];
    return 0;
}
```

Our C function callback simply forwards the message on to the main layer and pushes the puck out from the corner.

Simulating 3D Lighting in 2D Space

To simulate realistic lighting on the mallets, we resorted to a classic trick. If we center the light on the table, we can fake the direction of the light by rotating the mallets based on their position relative to the center (as shown in Figure 5-6). This will make the mallets' highlights and shadows appear to respond realistically to the light.

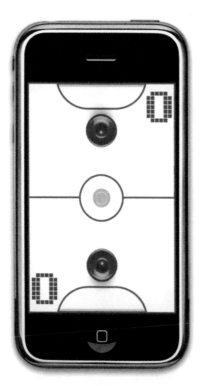

Figure 5-6. The minigolf game layer with player sprites rotated based on their angles to the center

In Listing 5-3, you'll learn how to rotate sprites relative to the center of the playing area to simulate a central light source.

Listing 5-3. *3D Lighting Simulation*

```
- (void)step: (ccTime) delta {

  // snip...

  // handle the rotation of the pucks to emulate
  // the effect of a light source
  cpVect newPosition;

  // find the center of the stage
  CGRect wins = [[Director sharedDirector] winSize];
  cpVect centerPoint = cpv(wins.size.width/2, wins.size.height/2);

  // rotate player 1
  newPosition = cpvsub(centerPoint, player1->p);
  [(Sprite*) p1_shape->data setRotation:90 -
    RADIANS_TO_DEGREES(cpvtoangle(newPosition))];

  // rotate player 2
  newPosition = cpvsub(centerPoint, player2->p);
  [(Sprite*)p2_shape->data setRotation: 90 -
    RADIANS_TO_DEGREES(cpvtoangle(newPosition))];
}
```

We first find the center of the stage by cutting the main window's width and height in half. For each player, we subtract the mallet's position from the center point and use that to calculate the angle with cpvtoangle, using the result to set the rotation of the sprite. We call the standard OpenGL RADIANS_TO_DEGREES macro to convert from radians returned by cpvtoangle into the degrees passed to setRotation.

Creating a Simple Application

In this section, we're going to build a simple prototype of a miniature golf game (shown in Figure 5-1). The user will drag a mallet with a finger and try to get the ball through the barriers to score a hole in one. Touching the back wall will cause the ball to return to the starting point.

Setting Up the Xcode Project

We start with a view-based application in Xcode (as shown in Figure 5-7), though we'll be trashing the view code and using the cocos2d Layer class for all of our graphics.

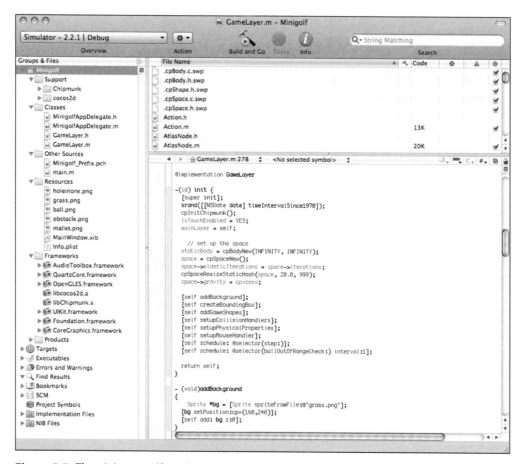

Figure 5-7. *The miniature golf Xcode project*

The application uses the following frameworks:

- *libcocos2d*: The 2D gaming library available at `http://cocos2d.org`. The version used in this project is 0.5.2; you may need to adjust your code if you use later versions, as the API is not yet stable.

- *libChipmunk*: The rigid body physics library at `http://code.google.com/p/chipmunk-physics/`.

- *AudioToolbox*: Makes the phone vibrate. (Yes, vibration is treated like any other sound.)

- *OpenGL ES*: A dependency of cocos2d. (See "Getting Started with Game Programming" for more details.)

- *CoreGraphics*: For drawing graphics on-screen and used by OpenGL ES.

- *QuartzCore*: Also for drawing graphics on-screen and used by OpenGL ES.

You'll need to drag the AudioToolbox, CoreGraphics, QuartzCore, and OpenGL ES frameworks to the Frameworks group in Xcode to start.

We also need to add the header and source files for cocos2d and Chipmunk to the Xcode project by dragging them to the Support group and to add some images to the Resources group for our different sprites.

Setting the Scene

We set the scene in the application delegate, a class that receives notifications from the application on major events like startup and shutdown. In this section, we'll define the scene and set things in motion for our main layer to start running.

Listing 5-4 shows how to set up the scene in the application delegate and hand off control to the Director.

Listing 5-4. *Setting Up the Scene in the Application Delegate*

```
- (void)applicationDidFinishLaunching:(UIApplication*)application
{
  [[Director sharedDirector] setAnimationInterval:1.0/60];
  Scene *scene = [Scene node];
  [scene add: [GameLayer node] z:0];
  [[Director sharedDirector] runScene: scene];
}
```

Now that we've set up the application delegate, we can get started on the main layer. The application starts in the MinigolfAppDelegate's applicationDidFinishLaunching: method. Here, we set the frame rate (60 frames per second in our example), create the main scene, and add a layer to it. Finally, we kick-start the cocos2d director into run mode by passing the scene we created.

```
- (void)applicationWillResignActive:(UIApplication *)application
{
  [[Director sharedDirector] pause];
}
```

```
- (void)applicationDidBecomeActive:(UIApplication *)application
{
  [[Director sharedDirector] resume];
}
```

If the user gets a call, we prepare our application to pause and resume sensibly by calling the relevant methods pause and resume on the Director singleton.

Creating the Game Layer

The game layer is where the action happens. This layer contains the walls of our space with a recessed area in the back, which will serve as the hole. We'll also add the ball, club, and some obstacles to make things interesting.

After the call to runScene (shown in Listing 5-4), our Layer subclass's init method is called (see Listing 5-5). This is where we create all our sprites, set them on the stage, and assign masses and frames to them so they'll act like real objects. We also set up all of the collision callbacks (C function pointers that will fire when objects collide), so we can respond to collision events between different types of objects. The collision types are defined in the enum at the top of *GameLayer.m*.

Listing 5-5. *Defining the Space in the Main Game Layer*

```
@implementation GameLayer

- (id) init {
    [super init];
    srand([[NSDate date] timeIntervalSince1970]);
    isTouchEnabled = YES;
    mainLayer = self;

    // set up the space for Chipmunk
    staticBody = cpBodyNew(INFINITY, INFINITY);
    space = cpSpaceNew();
    space->elasticIterations = space->iterations;
    cpSpaceResizeStaticHash(space, 20.0, 999);
    space->gravity = cpvzero;

    [self addBackground];
    [self createBoundingBox];
    [self addGameShapes];
    [self setupCollisionHandlers];
    [self setupPhysicalProperties];
    [self setupMouseHandler];
    [self schedule: @selector(step:)];
    [self schedule: @selector(ballOutOfRangeCheck:) interval:1];

    return self;
}
```

We start by seeding the random number generator and initializing Chipmunk (shown in Listing 5-5) and then create a space and enable bounce. We set the gravity to zero, since the field is flat and objects shouldn't be drawn in any direction by default.

```
- (void)addBackground
{
  Sprite *bg = [Sprite spriteFromFile:@"grass.png"];
  [bg setPosition:cpv(160,240)];
  [self add: bg z:0];
}
```

The background is just a sprite created from a resource in the Xcode project. We add it at the lowest z-index.

```
#define GOAL_MARGIN 145

// snip...

- (void)createBoundingBox
{
  cpShape *shape;

  CGRect wins = [[Director sharedDirector] winSize];
  startPoint = cpv(160,120);

  // make bounding box
  cpFloat top = wins.size.height;
  cpFloat WIDTH_MINUS_MARGIN = wins.size.width - GOAL_MARGIN;

  // bottom
  shape = cpSegmentShapeNew(staticBody, cpv(0,0),
    cpv(wins.size.width,0), 0.0f);
  shape->e = 1.0; shape->u = 1.0;
  cpSpaceAddStaticShape(space, shape);

  // top
  shape = cpSegmentShapeNew(staticBody, cpv(0,top),
  cpv(GOAL_MARGIN ,top), 0.0f);
  shape->e = 1.0; shape->u = 1.0;
  cpSpaceAddStaticShape(space, shape);
  shape -> collision_type = kColl_Horizontal;

  // and so on...
```

We then set up the bounding box, creating a small box behind the far wall, which we'll be using as our hole. The shapes are given collision types of kColl_Horizontal and kColl_Goal, which will later be hooked into the hole in one callback.

```
- (void)addGameShapes
{
  ball = [self addSpriteNamed:@"ball.png" x:160 y:120 type:kColl_Ball];
  obstacle1 = [self addSpriteNamed:@"obstacle.png" x:80
```

```
    y:240 type:kColl_Ball];
  obstacle2 = [self addSpriteNamed:@"obstacle.png" x:160
    y:240 type:kColl_Ball];
  obstacle3 = [self addSpriteNamed:@"obstacle.png" x:240
    y:240 type:kColl_Ball];
  player = [self addSpriteNamed:@"mallet.png" x:160
    y:50 type: kColl_Player];
}
```

To add the shapes, we extract out a convenience function, which sets up the sprite and its physical properties.

```
- (cpBody *) addSpriteNamed: (NSString *)name x: (float)x
    y:(float)y type:(unsigned int) type {

  UIImage *image = [UIImage imageNamed:name];
  Sprite *sprite = [Sprite spriteFromFile:name];
  [self add: sprite z:2];
  sprite.position = cpv(x,y);
```

We start by grabbing the image, creating and positioning its sprite to the x and y values passed in as arguments.

```
  int num_vertices = 4;
  cpVect verts[] = {
    cpv([image size].width/2 * -1, [image size].height/2 * -1),
    cpv([image size].width/2 * -1, [image size].height/2),
    cpv([image size].width/2, [image size].height/2),
    cpv([image size].width/2, [image size].height/2 * -1)
  };

  // all objects need a body
  cpBody *body = cpBodyNew(1.0, cpMomentForPoly(1.0, num_vertices,
    verts, cpvzero));
  body->p = cpv(x, y);
  cpSpaceAddBody(space, body);

  // as well as a shape to represent their collision box
  cpShape* shape = cpCircleShapeNew(body, [image size].width / 2,
    cpvzero);
  shape->data = sprite;
  shape -> collision_type = type;

  if (type == kColl_Ball) {
    shape->e = 0.5f; // elasticity
    shape->u = 1.0f; // friction
  } else {
```

```
    shape->e = 0.5; // elasticity
    shape->u = 0.5; // friction
  }
```

We then give it a body and a shape, assign elasticity and friction, and set the shape's data and collision type. If you want different values for different shapes, you'll want to set that here as it's difficult to get at the shape later.

```
  cpSpaceAddShape(space, shape);
  return body;
}
```

Finally, we add the shape to the space before returning the body.

```
// collision types
enum {
  kColl_Ball,
  kColl_Goal,
  kColl_Horizontal,
  kColl_Player
};


// snip...


- (void)setupCollisionHandlers
{
  cpSpaceAddCollisionPairFunc(space, kColl_Ball, kColl_Goal,
    &holeInOne, ball);
  cpSpaceAddCollisionPairFunc(space, kColl_Ball, kColl_Horizontal,
    &restart, ball);
}
```

Setting up the collision handlers is just a matter of calling cpSpaceAddCollisionPairFunc with our space, the two types colliding, a pointer to the function we want to call when they collide, and some data to be passed to the callback (we won't be using this, but you have to send something).

```
void resetPosition(cpBody *ball) {
  cpBodyResetForces(ball);
  ball -> v = cpvzero;
  ball -> f = cpvzero;
  ball -> t = 0;
  ball -> p = startPoint;
  AudioServicesPlaySystemSound(kSystemSoundID_Vibrate);
}
```

```
static int holeInOne(cpShape *a, cpShape *b, cpContact *contacts,
  int numContacts, cpFloat normal_coef, void *data) {
  GameLayer *gameLayer = (GameLayer *) mainLayer;
  [gameLayer holeInOne];
  return 0;
}

static int restart(cpShape *a, cpShape *b, cpContact *contacts,
  int numContacts, cpFloat normal_coef, void *data) {
  cpBody *ball = (cpBody*) data;
  resetPosition(ball);
  return 0;
}
```

The callback functions themselves are straightforward. A hole in one forwards the message to the GameLayer, where victory is signaled with a sprite overlaid on top of the main layer and a quick vibration of the phone.

```
- (void)setupPhysicalProperties
{
  cpBodySetMass(ball, 25);
  cpBodySetMass(obstacle1, INFINITY);
  cpBodySetMass(obstacle2, INFINITY);
  cpBodySetMass(obstacle3, INFINITY);
  cpBodySetMass(player, 2000);
}
```

We set the mass of the player to be substantially larger than the mass of the ball to get a good speed when hitting and set the obstacles to INFINITY so that they don't move when hit (they are anchored to the ground after all).

```
- (void)setupMouseHandler
{
  playerMouse = cpMouseNew(space);
  playerMouse->body->p = player->p;
  playerMouse->grabbedBody = player;

  // create two joints so the body isn't rotated
  // around the finger point
  playerMouse->joint1 = cpPivotJointNew(playerMouse->body,
    playerMouse->grabbedBody,
    cpv(playerMouse->body->p.x - 1.0f,
    playerMouse->body->p.y));
  cpSpaceAddJoint(playerMouse->space, playerMouse->joint1);

  playerMouse->joint2 = cpPivotJointNew(playerMouse->body,
    playerMouse->grabbedBody,
```

```
      cpv(playerMouse->body->p.x + 1.0f,
      playerMouse->body->p.y));
   cpSpaceAddJoint(playerMouse->space, playerMouse->joint2);
}
```

Next, we set up the mouse handler. This lets us treat the mallet as a draggable cursor while
retaining its physical properties when it interacts with the other objects. We give it two
joints (points which serve as axes for rotation in the mouse), because we'll be positioning it
in front of the finger, and we don't want it to rotate around the touch point when moved.

```
- (void)touchesMoved:(NSSet*)touches withEvent:(UIEvent*)event{
   CGPoint playerTouchLocation = CGPointMake(-300, 240);

   for (UITouch *myTouch in touches) {
     CGPoint location = [myTouch locationInView: [myTouch view]];
     location = [[Director sharedDirector] convertCoordinate: location];
     // set the finger location to be the lowest touch
     playerTouchLocation.x = location.x;
     playerTouchLocation.y = location.y;
   }

   // into game coords...
   CGPoint location = playerTouchLocation;
   cpFloat padding = finger_padding * ((120 - location.y) / 100);
   location.y -= padding;
   location.y += fat_fingers_offset;
```

In the touch event handler, we grab the last touch in the set (in this case, the only touch) and
compensate for the user's finger based on how far up it is on the field. This way, the user can
see the mallet and still bring the mallet all the way back to the beginning of the field.

```
   // trap the location to half-field
   if (location.y > 230) location.y = 230;
   if (location.y < 0) location.y = 0;
```

We trap the position to half-field so that the user can't cheat.

```
   cpVect playerposition = cpv(location.x, location.y);
   cpMouseMove(playerMouse, playerposition);
}
```

And finally, we create a vector and move the mouse to it.

```
- (void)touchesBegan:(NSSet *)touches withEvent:(UIEvent *)event {
   [self touchesMoved:touches withEvent:event];
}
```

The touchesBegan: withEvent: handler is identical to touchesBegan: withEvent: but could be different depending on what you want to trigger when dragging starts.

Finally, we schedule the following two callbacks:

```
static void eachShape(void *ptr, void* unused) {
  cpShape *shape = (cpShape*) ptr;
  Sprite *sprite = shape->data;
  if (sprite) {
    cpBody *body = shape->body;
    [sprite setPosition: cpv(body->p.x, body->p.y)];
  }
}

// snip...

- (void)step: (ccTime) delta {
  int steps = 1;
  cpFloat dt = delta/(cpFloat)steps;

  for (int i=0; i<steps; i++) {
    cpSpaceStep(space, dt);
  }

  cpSpaceHashEach(space->activeShapes, &eachShape, nil);
  cpSpaceHashEach(space->staticShapes, &eachShape, nil);

}
```

The first is a C function, eachShape, which makes the animation engine run. It starts a timer that calls step: on our layer at regular intervals. In the step: function, we increment the physical properties of the layer by calling cpSpaceStep with the time difference. We step the display by sending cpSpaceHashEach for the active and static shapes, passing it a function that positions the sprites.

```
- (void) ballOutOfRangeCheck: (ccTime) delta {
  if (ball -> p.x > 320 || ball -> p.x < -80 ||
    kball -> p.y > 550 || ball -> p.y < -80) {
      resetPosition(ball);
    }
}
```

The function ballOutOfRangeCheck: checks to see if the ball has been pushed so fast that it passed through the wall. If so, we treat it as having gone off the course in the same way we treated the puck going off the hockey table, and we reset the position for the user to try again.

```
- (void)holeInOne
{
  AudioServicesPlaySystemSound(kSystemSoundID_Vibrate);
  holeInOneBG = [Sprite spriteFromFile:@"holeinone.png"];
  [holeInOneBG setPosition:cpv(160,240)];
  [self add:holeInOneBG z:10];
  [self performSelector:@selector(resetGame:) withObject:nil
    afterDelay:2.0];
}

- (void)resetGame:(id)object
{
  [self remove:holeInOneBG];
  resetPosition(ball);
}
```

Finally, we include the function holeInOne, which is called when the user scores. Vibration is done with AudioServicesPlaySystemSound using the kSystemSoundID_Vibrate constant. Finally, the function sets a timer to remove the message and reset the game after 2 seconds.

Summary

You now have enough knowledge to start messing around with the laws of nature.

We've really only begun to scratch the surface of what can be done with Chipmunk and cocos2d. You can create skeletons and joint systems, which can be snapped together like LEGO bricks to simulate anything from a simple bicycle to a human being. You can then throw your objects into a space with whatever goals and constraints you choose to define to make the game a challenge. If you want to make things even more difficult, you can play with gravity, or even use the accelerometer to alter the gravity of the scene. Of course, as with any decision you make in your code, you should always be questioning whether or not you're adding to the playability of the game.

Neil Mix

Company: *Pandora Media, Inc.*

Location: *Appleton, Wisconsin*

Former life as a developer: Building Internet applications using languages including JavaScript, Objective-C, C, Java, Perl, C++, and SQL. User interaction designer, software architect, and programming language theory wannabe.

Life as an iPhone developer: Built the Pandora Radio music application using Xcode

What's in this chapter: This chapter explores basic audio development for the iPhone, including an introduction to Core Audio, audio streaming and networking, architecture for audio applications, and advanced streaming techniques for the mobile environment.

Key technologies:

- *Core Audio*

- *NSURLConnection*

- `AudioFileStream`

- `AudioQueue`

Chapter 6

Serious Streaming Audio the Pandora Radio Way

This chapter will walk you, step by step, through the process of downloading and playing audio on the iPhone using Pandora Radio, the top application for the iPhone in 2008. By the end of this chapter, you will have learned about Apple's audio API infrastructure (known as Core Audio), how to turn bytes sent over HTTP into playable audio, and the challenges unique to delivering audio content in a mobile environment.

Choosing to Develop for the iPhone

Pandora is a popular Internet radio music service that's been publicly available on the Web since 1995. One of Pandora's core goals, however, is to bring music to listeners wherever they are, including when they are away from their computers. So it was no surprise when Pandora expanded its service to include mobile phones in 1996. Unfortunately, early Pandora mobile implementations didn't attain popularity as quickly as Pandora had on the Web. But when the iPhone arrived and the public SDK for iPhone became a reality, we at Pandora were eager for the opportunity to deliver our service to this great new device.

Pandora and the iPhone provide an ideal fit for each other. Prior to the iPhone, many people in the United States didn't regard mobile phones as entertainment devices. But the iPhone has changed that, unleashing new opportunities for portable entertainment wherever you go. And the iPhone's focus on user experience (another core value at Pandora as well) meant we could deliver our service with a great user interface (UI).

I've primarily been a web hacker throughout my career, working mostly in scripting languages and dabbling in statically typed compiled languages for only short periods of time. So how did I cross over to C and Objective-C as required for iPhone development? By jumping in headfirst.

I knew a little C (who doesn't?) prior to building Pandora's iPhone application but nothing of Objective-C or the Cocoa way. I had to learn in transit, sprinting to get our application ready in time for the App Store launch. It reminded me of the time I went tandem skydiving: I was taught the mechanics of jumping while we were still on the ground, but I was taught how to land while floating down in the parachute! And just like parachute-landing skills can be bootstrapped during skydiving, much of my iPhone audio knowledge was bootstrapped during the process of building our application.

I regularly hear from programmers who've fallen into the trap of believing that everyone knows more about iPhone programming than they do. But the truth is, we're all learning as we go. Despite the fact that I'm a developer of one of the top iPhone applications, I'm still in the process of learning how to develop for this platform. There's no magic bullet here, no secret knowledge. It's just hard, painstaking work.

Introducing Pandora Radio's Technology

Pandora Radio is renowned for having a great UI. But a polished UI alone won't make an application great. The mobile environment presents incredible challenges for providing Internet audio. And thus the hidden gem of the Pandora Radio application is its ability to deliver seamless, high-quality audio to your phone wherever you are, regardless of connection type—even over cellular networks such as EDGE.

In order to provide a terrific audio experience, though, you must first know the tools that are required. It's also useful to practice good habits in your programming.

Grasping the Basics of Audio Development

Writing software that plays audio is hard, especially as you get closer to the hardware layer. Apple's Core Audio API does a lot to abstract out the hardware specifics, but it's close enough to the hardware to make coding to the API a challenge. In addition, debugging broken audio is hard: how do you debug silence? The audio output alone doesn't provide many clues as to what's going wrong.

Since Core Audio can be so challenging and debugging so hard, you'll do yourself a favor if you get some audio—any audio!—playing in the simplest possible way before moving on to more advanced functionality. That way, if you make changes along the way that break your audio, you always have a working example to revert to, which can help you figure out what went awry.

Also, it's very important that you *always* check the return codes from audio API calls. Even if you don't do any specific error handling, you should at least log any errors. Having these log entries helps a ton when debugging mysterious audio problems.

TIP

Start with a simple, working implementation, and check all error codes.

This chapter's sample application will demonstrate these principles: we're going to create the simplest possible working implementation first, and at the end of this chapter, we'll discuss how to improve it.

Also, as you browse the source code, you'll see that it carefully checks the status codes of each Core Audio API call like this:

```
if (VERIFY_STATUS(status)) { ...
```

We may not always check the return value of VERIFY_STATUS, but we almost always call VERIFY_STATUS when we use a Core Audio API. To see why this helps, let's trace VERIFY_STATUS back to its definition:

```
#define VERIFY_STATUS(status) \
        AudioPlayerVerifyStatus(status, __FILE__, __LINE__)
```

Notice the __FILE__ and __LINE__ macros. These allow us to capture the file name and line number at which the macro is called, which is useful when debugging. Next, we'll trace this a step further to the AudioPlayerVerifyStatus definition:

```
BOOL AudioPlayerVerifyStatus(OSStatus status, char *file, int line) {
  if (status == noErr) {
    // Logging successes is prolific but useful for debugging.
    // We'll turn it off by default, but if you encounter a problem, you
    // can uncomment this to trace the path of execution.
    //NSLog(@"success at %s:%i", file, line);
  } else {
    char *s = (char *)&status;
    NSLog(@"error number: %i error code: %c%c%c%c at %s:%i",
          status, s[3], s[2], s[1], s[0], file, line);
  }
  return status == noErr;
}
```

You see that we log the error code and its location whenever an error occurs. If, during the course of building and debugging your application, you encounter problems with audio, the logs will give you an immediate indication of what went wrong and where in your Core

Audio calls, even if you don't explicitly handle error cases. This technique can be a valuable time saver.

Managing Complexity

As your application grows and you add features and fix bugs, your audio code will quickly become complex. Complexity is inevitable and only grows with time, so be sure to leave time for refactoring and code simplification as you add features. Don't worry about performance issues until you encounter problems. The iPhone is surprisingly robust, and you may never encounter the performance problems you expect.

For example, it's common for applications to perform audio functions in a separate audio thread to keep the application robust and responsive. But with the Pandora Radio application, we've avoided threading and found negligible impact on application performance. The simplification gained by not having to deal with threading complexity has been a great time saver, which has allowed us to add great new features more quickly than we otherwise would have. We may have to put audio in a separate thread eventually, but in the meantime, we gain more by keeping the code simple. (It's also worth remembering that, since the iPhone has only one single-core ARM processor, threading is not as beneficial as in desktop environments.)

TIP

> Stay focused on code simplicity, and don't optimize before you need to.

Outlining Our Sample Application

The application we're going to build in this chapter is a simple music player. Its UI will consist of a text field for the URL of an audio file and buttons to play or pause the audio. Since we're focusing on audio in this chapter, we won't spend time walking through the UI, so we can focus solely on the audio code.

Apple's audio APIs are known as Core Audio and consist of a toolbox of components that can be used to slice audio files into pieces and stream them to the audio hardware. One thing that is *not* a part of Core Audio is networking, and keeping this in mind is important as you design your code. Despite the fact that we're playing audio delivered over a network, audio transport is a distinct task from audio playback. If you are able to keep these tasks clearly delineated in code, you'll have an easier time refactoring and adding features.

Streaming Audio

When delivering audio to a device via the Internet, you can employ two models of delivery: streaming or downloading.

With streaming audio, the audio servers deliver bytes over the network at the same bit rate as the audio. For example, audio encoded as 64kbps would be streamed out at a rate of, you guessed it, 64kbps. Streamed audio is typically used when you have a continuous, unending music stream.

In the download model of audio delivery, your music is divided into discrete chunks of data (for example, one song at a time), and the client will download one data chunk as fast as possible, play it, and begin downloading the next chunk only when the first chunk is nearing completion.

Each model has advantages and disadvantages. Streaming requires less memory overhead on the client (data is consumed as soon as it is delivered) and requires less complexity for the client implementation (at the expense of more complexity on the server). But streaming can also be somewhat vulnerable to audio disruptions due to network latency and dropouts. Downloaded audio, on the other hand, has greater memory overhead (since the audio segments may be quite large) and more client complexity. But downloaded audio is also more resilient to network disruptions, since the network bandwidth is fully utilized rather than capped at a fixed amount.

There's no right or wrong answer in choosing between streaming or downloading, and different services use different models. Pandora happens to use a downloading model, which is the model we'll use for our sample application as well. The download model makes our servers simpler and our audio less prone to interruptions. Your mileage may vary.

Our application will download audio by requesting files over a simple HTTP connection. To do this on the iPhone, you may use NSURLConnection, an Objective-C API, or it's plain C counterpart CFURLRequest. The Core Audio documentation tends to provide examples that use CFURL, and CFURL is a rawer API, with more configuration capability than NSURLConnection. However, we've chosen to use NSURLConnection in Pandora Radio, because it's easier to use than CFURLRequest and provides plenty of power for our purposes. So we'll be using NSURLConnection in our sample application.

Once bytes have been received over the network, we need to hand them off to Core Audio for playback. Figure 6-1 shows how this process works: bytes are received from the network, and bytes are then parsed into audio packets using AudioFileStream. Buffers are allocated via AudioQueue to hold packets, and then buffers are sent to the audio hardware using (again) the AudioQueue. While "parsing" and "hardware" may sound a bit scary and complex, the Core Audio APIs hide most of the difficult details of this process.

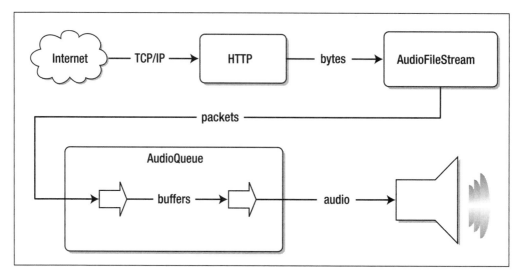

Figure 6-1. *Transferring Internet audio data to output hardware on the iPhone*

Keeping Your Code Format Agnostic

Core Audio was designed with the goal of complete agnosticism across all file formats and audio codecs. In other words, the goal is that you should be able to use the same code for any supported type of audio.

To accomplish this, Core Audio applies certain generalizations to all audio types. All audio files are treated as a series of audio packet data that are enclosed in an envelope. The encoding format of the packet data is distinct from the envelope format, so you can mix and match to some extent.

It works like this: AudioFileStream reads enough of the incoming data to determine the type of the file and its overall structure. Once it has done this, it provides an AudioStreamBasicDescription struct to you, and you then pass this struct to your AudioQueue so that it knows what kind of data it will be decoding. From there, the AudioFileStream will produce packets of audio data. You then use AudioQueue to allocate buffers that will store the packets as they are queued up for delivery to the audio hardware.

Using Envelopes and Encoding

As stated previously, the parsing and queuing process works the same way for all supported audio types. For example, you're probably familiar with Advance Audio Coding (AAC) audio files. What you may not know is that AAC is a way to encode audio data, but there are multiple envelope formats that can deliver AAC encoded audio. One such commonly known format is MPEG-4 Part 14 (referred to as MP4, also known as M4A), the format that iTunes uses by default. M4A is the envelope format, and AAC is the encoding.

As it turns out, M4A isn't the best format for streaming audio. It contains a significant amount of data at the beginning defining the structure of the file, the presence of which can increase delay to the first music you hear. It's also difficult to begin playback in the middle of an M4A file. A better envelope for delivering audio is Audio Data Transport Stream (ADTS), a format that includes structural playback data all throughout the file rather than at the beginning. Thus, ADTS can play audio with a smaller up-front data size and more easily support playback of audio starting in the middle of a file (a feature that Pandora Radio uses to great effect).

While the Core Audio API insulates you from having to know these details, it will nevertheless be important for you to understand them if you intend to deliver audio to the iPhone. Which format produces the smallest files? What are the licensing restrictions? Which encoding rate produces the best trade-off of audio quality to file size?

Designing Our Sample Application

Now that you have a basic understanding of the technologies we'll use to build our application, it's time to think about how we'll implement it. As a first step, we'll look at the modules we'll create and how they fit together.

To create an application that uses audio, you create a project as normal in Xcode: select **File ➤ New Project** from the Xcode menu, and make sure the Application item under iPhone OS is selected. Next, choose the type of application you wish to create (in this case, View-Based Application), and select Choose.

Once you've created your project, make sure to add the AudioToolbox framework: control-click **Frameworks** in the **Groups & Files** sidebar on the left, and select **Add ➤ Existing Frameworks**, as shown in Figure 6-2.

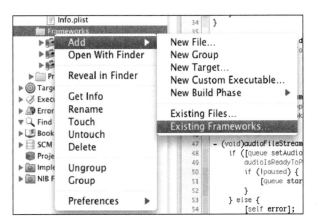

Figure 6-2. *How to add a framework to your Xcode project*

As always, when choosing the AudioToolbox framework, make sure you're selecting from the correct SDK directory path. In the case of our sample application, the framework is here: */Developer/Platforms/iPhoneOS.platform/Developer/SDKs/iPhoneOS2.0.sdk/System/Library/ Frameworks/AudioToolbox.framework.*

Choosing *iPhoneOS2.0.sdk* allows for the application to be the most broadly compatible and therefore installable on the largest number of devices. If you are using features that only exist on newer versions of the iPhone operating system, you should select frameworks from the corresponding SDK.

The complete sample project in Xcode would look like Figure 6-3.

Figure 6-3. *Our sample application's Xcode project*

There are four Objective-C classes we'll create to stitch our application's audio together:

- `AudioRequest`: This class is responsible for initiating an `NSURLConnection` to download our audio and deliver raw bytes to our `AudioFileStream`.

- `AudioFileStream`: This is a simple Objective-C wrapper around the plain C `AudioFileStream` class that receives raw data from the network and parses it into audio packets.

- `AudioQueue`: This is a simple Objective-C wrapper around the plain C `AudioQueue` class that receives audio packets from our `AudioFileStream`, converts them into buffers, and then sends the buffers to the audio hardware.

- `AudioPlayer`: This could be considered the commander class that is responsible for controlling audio playback and reporting playback status. `AudioPlayer` coordinates the transfer of raw audio data from `AudioFileRequest` into `AudioFileStream` packets, then into `AudioQueue` buffers, and finally onto the audio hardware.

How this process works will become clear as we look at the code. In the meantime, Figure 6-4 provides a bird's eye view of how audio data flows through the various classes.

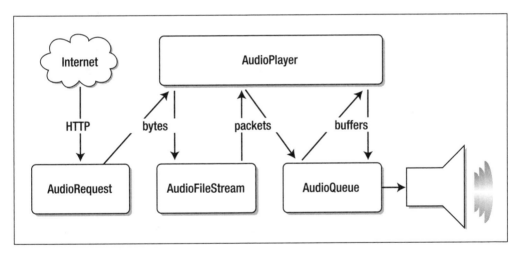

Figure 6-4. *Audio data flowing through our classes*

Implementing the Player

Now that you understand the components of our application, let's jump right into its implementation.

As we proceed into implementation of our application, be mindful of earlier tips. For one, we're going to avoid using threads (as mentioned earlier, at the time of this writing, the Pandora Radio application uses only one thread for application code). Second, we'll make sure to fastidiously handle any errors from the Core Audio API.

Also, we want to make sure our application plays nicely with the iPhone generally. This is no small feat, and for audio applications, the first place to start is by using AudioSession.

AudioSession

AudioSession manages the audio profile of our application and helps us cleanly handle interruptions such as incoming phone calls or text messages. If you open *AudioPlayerAppDelegate.m* and look at the applicationDidFinishLaunching: method, you'll see something like the following:

```
AudioSessionInitialize (NULL, NULL, interruptionListener, self);
```

The applicationDidFinishLaunching: message is sent to a UIApplicationDelegate exactly once on application startup and is the ideal location for application initialization such as AudioSessionInitialize. This declaration of our audio session is the first step in guaranteeing our application plays nicely with the audio subsystem used by all applications on the phone.

Take a moment to imagine what it takes to play audio on the iPhone. There are several different audio routes—incoming audio through the microphone or headphones and outgoing audio through the ear speaker, the external speaker, or the headphones. In addition, audio played by an application may be interrupted at any time, say, for a phone call or text message. Plus, audio may be interrupted when the user presses the microphone button that plays audio via the iPod application. Also, different applications have different sound needs, and for some applications (such as Pandora Radio), it would be entirely inappropriate for the application to halt when the phone is locked. Then too, the volume controls need to adjust both ringer volume and media volume based on context. And to top it all off, the hardware capacity for audio is limited and is often unable to play audio from multiple sources simultaneously. That's a lot to manage!

To help manage all these scenarios, the Core Audio experts invented the concept of an AudioSession, which tells the operating system what our audio needs are and lets it tell us what its audio needs are. The AudioSessionInitialize call establishes a session and registers a callback method with the operating system to be executed when our application's audio is interrupted due to phone call, for example.

"C has callback functions?" you may ask. The C programming language allows you to pass functions as parameters to other functions (much like passing a selector in Objective-C). You're not really passing a function but rather the address of a function, and this mechanism is great for writing event-based APIs in C. You write a C function that performs audio interruption handling (in this case, we named it interruptionListener and defined it at the end of the file) and pass a pointer to that function into AudioSessionInitialize. When an audio interruption occurs, your function is called back just like any normal C function.

Following the audio session initialization, you'll see these lines:

```
UInt32 sessionCategory = kAudioSessionCategory_MediaPlayback;
AudioSessionSetProperty (
    kAudioSessionProperty_AudioCategory,
    sizeof (sessionCategory),
    &sessionCategory);
```

This declares our session as being a "media playback" application. Why is this important? First of all, this declaration alters the behavior of the volume controls so that they apply to media volume, not ringer volume. Second, it alters the behavior of the device when locked. Normally, the device goes into a hibernation mode some time after the device has been locked. This hibernation results in the halting of an application: its run loops are suspended, and no events are processed. If this happened to a media playback application such as Pandora Radio, the playback would eventually halt. By declaring our session as MediaPlayback, the operating system will know to keep our application running when the device is locked, so we can continue playing music indefinitely.

So you can see there are benefits to correctly declaring your audio session. But it's also important to realize that audio sessions start out inactive and must be explicitly activated to have the desired effects. It is best practice to activate your audio session only when audio is playing so that the device may hibernate when locked. This helps lengthen battery life. If you find the load: method of AudioPlayer, you see that we activate our audio session by calling AudioSessionSetActive(YES). And when audio completes in audioPlayerPlaybackFinished:, we deactivate the session by calling AudioSessionSetActive(NO) so that the phone may hibernate.

When we receive an incoming call during playback, our interruptionListener function will be executed. It is very important that you handle interruptions cleanly. If you don't, you may crash your application or even the phone. So please, handle the interruptions, and *test your application* by calling your phone during playback. Interruption handling is tricky and prone to bugs. This will be the most buggy and least tested portion of your application and is a key feature for integrating nicely with a mobile device.

AudioRequest

The AudioRequest class wraps the behavior of NSURLConnection and simplifies its delegate implementation to the following two messages:

```
- (void)audioRequest:(AudioRequest *)request didReceiveData:(NSData *)data;
- (void)audioRequestDidFinish:(AudioRequest *)request;
```

The first message lets you respond to the receipt of incoming data, which we'll eventually pass on to AudioFileStream. The second message lets us clean up when the connection is complete.

But what you don't see here may cause you to raise an eyebrow with concern. Where's the error handling? Surely we don't assume that all connections succeed!

Of course we don't. We just handle connection errors at a different location in code. Network errors aren't the only type of errors that can happen when playing audio, which means that AudioRequest isn't the right location to enforce error-handling behavior. For example, the file you download might contain an audio format your device can't handle, or might not be an audio file at all. You'll handle that kind of error downstream from AudioRequest, so why make things more complicated? Connection errors and file format errors can all be handled in the same way, downstream from AudioRequest. Looking at Figure 6-4, it's easy to see how the AudioPlayer class is a great place for detecting and communicating such errors to a higher level delegate. (Although AudioRequest should still log any network errors to help with debugging! Handling an error downstream is entirely different than completely ignoring it.)

"But what if the connection fails midway through the download?" you might ask. "That's not the same as a file format error."

That's a terrific point which begs the question, what should you do if a connection fails midway through download, after audio has started playing? Ideally, you'd try to heal the connection, but that's a slightly more complicated feature we'll discuss in the "Dropped Connections" section at the end of this chapter. For now, we'll just log the error (for debugging) and end the song gracefully. Ultimately, your listeners will understand that something didn't go as planned, so there's no reason to pester them with cryptic error messages too. Remember, it is likely that your listener has the device in a pocket with the screen locked at the moment this happens. Error messages wouldn't be seen until long after they occur, which can be confusing.

The implementation of AudioRequest is pretty straightforward. When the application receives bytes, it forwards them onto the delegate. When the connection completes or errors out, it notifies the delegate of completion.

The trick here comes in the handling of connection:didReceiveResponse: messages from NSURLConnection—this message is called for both HTTP success *and* failure responses. You *might* expect connection:didFailWithError: to notify of HTTP error responses, but it doesn't. NSURLConnection is designed for any type of protocol, not just HTTP. So connection:didFailWithError: indicates a *connection* failure, not a protocol failure.

For example, a TCP/IP networking error could cause connection:didFailWithError: to be called at any time; it could be called before any connection is established (in which case a connection will never be established), or midway through audio download. On the other hand connection:didReceiveResponse: is called immediately after a connection is established and prior to any data transmission. An HTTP error such as 404 (File Not Found) would cause a connection to be established as normal, and thus connection:didFailWithError:

would never be called. Instead, `connection:didReceiveResponse:` is called with an NSHTTPURLResponse object that indicates the error.

Fortunately, handling HTTP protocol errors is pretty easy (especially given our approach toward error handling):

```
- (void)connection:(NSURLConnection *)connection
    didReceiveResponse:(NSURLResponse *)aResponse
{
    if ([aResponse isKindOfClass:[NSHTTPURLResponse class]]) {
        NSHTTPURLResponse *response = (NSHTTPURLResponse *)aResponse;
        if (response.statusCode >= 400) {
            NSLog(@"AudioRequest error status code: %i",
                    response.statusCode);
            [delegate audioRequestDidFinish:self];

            // Prevent further receipt of data that are not audio
            // bytes and therefore could harm the audio subsystems.
            [self cancel];
        }
    }
}
```

AudioFileStream

Core Audio is a C API, which gives it flexibility to be used both in Objective-C and C++ projects. But C APIs aren't always the friendliest to deal with. Since iPhone application coding requires Objective-C, we can wrap `AudioFileStream` in an Objective-C wrapper to make it cleaner, easier to read, and more modular with respect to our other classes. The benefit of this will become clear when we get to the `AudioPlayer` class.

One thing that makes `AudioFileStream` a bit unwieldy is its use of function pointers and callback functions. For example, here we open a file stream:

```
AudioFileStreamOpen(self, propertyCallback, packetCallback, 0, &streamID);
```

The first parameter is `userData`, which is passed to all callback events from the stream. Next, you'll see the `propertyCallback` and `packetCallback` parameters. Those are pointers to C functions with specific signatures. The `userData` (our `self` object) is passed into these callbacks, which allows the callbacks to maintain context despite their asynchronous nature.

This API design works well for a C API, but function pointers are pretty rare in Objective-C APIs. The Objective-C way of doing this is to have a delegate: our `AudioFileStream` class wraps these callback functions and turns them into delegate responses. For example, here's the definition of our `packetCallback` function:

```
void packetCallback(
    void *clientData,
    UInt32 byteCount,
    UInt32 packetCount,
    const void *inputData,
    AudioStreamPacketDescription *packetDescriptions)
{
    AudioFileStream *self = (AudioFileStream *)clientData;
    [self didProducePackets:packetDescriptions
        withPacketCount:packetCount
        fromData:inputData
        andByteCount:byteCount];
}
```

The callback is very simple: it grabs our `self` object (the `userData` we passed in when opening the file stream) and forwards the incoming parameters onto `self` in a message. That
message in turn passes the data to our delegate:

```
- (void)didProducePackets:(AudioStreamPacketDescription *)desc
    withPacketCount:(UInt32)packetCount
    fromData:(const void *)inputData
    andByteCount:(UInt32)byteCount
{
    [delegate audioFileStream:self
        didProducePackets:[NSData dataWithBytes:inputData length:byteCount]
        withCount:packetCount
        andDescriptions:desc];
}
```

It's a simple technique, but makes a big difference in readability when we get to the
`AudioPlayer` class, which receives this delegate message.

One other reason to wrap `AudioFileStream` in an Objective-C wrapper is to narrow the
API for our specific purposes. `AudioFileStream` contains a lot of functionality we don't
use, so our wrapper can improve readability by not exposing those pieces. For example, our
`propertyCallback` function receives notifications about many different properties, but
we're only interested in a couple of them. By wrapping `AudioFileStream` in a wrapper, we
simplify the API that its consumer must know and respond to.

You may also notice that `AudioFileStream` contains a strange thing called a magic cookie.
The **magic cookie** is an abstract concept introduced by the Core Audio API to represent
data specific to a particular audio format that is required for decoding. It's an opaque
structure for which you don't need to know specific detail. It's just important to know that
`AudioFileStream` *may* generate a magic cookie, and if it does, it needs to be passed onto
`AudioQueue` to enable proper decoding.

AudioQueue

Much like AudioFileStream, the AudioQueue C API is fairly large and sometimes complicated. We wrap it in an Objective-C wrapper to condense the API into a more manageable
bite-sized chunk.

There's not a whole lot to say about our AudioQueue class without digging deep into the
details of implementation. You'll see there are public methods for setting the stream description and magic cookie (which intentionally correspond directly to delegate events sent by
AudioFileStream), methods to start and pause the queue, methods for buffer allocation
and queuing, and an end-of-data notification method. Take a look through *AudioQueue.m*
for the nitty-gritty details regarding how to use the AudioQueue C API.

AudioPlayer

AudioPlayer is the workhorse class that stitches the AudioRequest, AudioFileStream, and
AudioQueue together. It's responsible for a lot of choreography but is surprisingly simple
thanks to our earlier efforts to simplify using wrapper classes. What follows is the key section
of code that demonstrates how audio data flows between application components (as dia-
gramed previously in Figure 6-4) and is the key to playing Internet audio in our application:

```
- (void)audioRequest:(AudioRequest *)request didReceiveData:(NSData *)data
{
    if ([fileStream parseBytes:data] != noErr) {
        [self error];
    }
}

- (void)audioFileStream:(AudioFileStream *)stream
    foundMagicCookie:(NSData *)cookie
{
    [queue setMagicCookie:cookie];
}

- (void)audioFileStream:(AudioFileStream *)stream
    isReadyToProducePacketsWithASBD:(AudioStreamBasicDescription *)absd
{
    if ([queue setAudioStreamBasicDesciption:absd] == noErr) {
        audioIsReadyToPlay = YES;
        if (!paused) {
            [queue start];
        }
    } else {
        [self error];
    }
}
```

```
- (void)audioFileStream:(AudioFileStream *)stream
    didProducePackets:(NSData *)packetData
    withCount:(UInt32)packetCount
    andDescriptions:(AudioStreamPacketDescription *)packetDescriptions
{
    AudioQueueBufferRef bufferRef;
    OSStatus status = [queue
        allocateBufferWithData:packetData
        packetCount:packetCount
        packetDescriptions:packetDescriptions
        outBufferRef:&bufferRef];
    if (status == noErr) {
        [queue enqueueBuffer:bufferRef];
    } else {
        [self error];
    }
}

- (void)audioQueuePlaybackIsStarting:(AudioQueue *)audioQueue {
    [delegate audioPlayerPlaybackStarted:self];
}

- (void)audioQueuePlaybackIsComplete:(AudioQueue *)audioQueue {
    [delegate audioPlayerPlaybackFinished:self];
}
```

The flow is straightforward and reflects what you saw in Figure 6-4: when we receive data from the network, we pass it to AudioFileStream. When AudioFileStream finds a stream description, magic cookie, and packet data, it passes them to AudioQueue. When AudioQueue starts playing audio and finishes playing audio, it lets our delegate know.

That's it! You've now got audio data flowing from the network down to the iPhone hardware.

Ending with a New Journey

You've now passed your first hurdle and have audio playing on the iPhone. Unfortunately, you have a few hurdles ahead before you've achieved world-class, robust, and reliable audio streaming. Hopefully, the simplified groundwork we've laid will let you pass these hurdles easily and quickly. Let's discuss what problems now await you.

Falling Behind in a Slow Network

One tricky aspect of AudioQueue is that it meticulously manages the delivery of audio with respect to time. This careful clock management enables complex audio synchronization tasks, but makes our job a little more difficult. The timeline of an AudioQueue continues moving forward even if there's no audio to play, so if an audio packet happens to be queued late, after it would have otherwise played, AudioQueue skips the packet and instead waits for packets that can play in sync with the timeline. In a slow network (such as an EDGE cellular network), your audio bit rate may be close to the maximum capacity of your network connection. Therefore, it's common for audio packets to be delayed awaiting incoming data from the network. But the time synchronization behavior of AudioQueue means that if you fall behind in these conditions, the maximum capacity of the network prevents you from making up for lost time, and *you may never catch up*. Figure 6-5 demonstrates this condition. This is an absolute disaster for user experience, where audio stops in a long, dead silence until the song is complete.

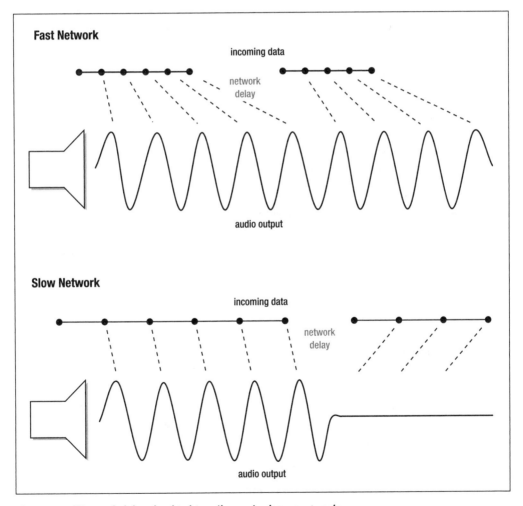

Figure 6-5. *Network delays lead to long silences in slower networks*

The solution to this problem is to pause the AudioQueue when you run out of data from the network. This is a little tricky, because you don't receive notification from AudioQueue when you run out of data. You instead have to infer based on the size of the packets you've queued and the `audioQueue:isDoneWithBuffer:` callbacks you've received. You also have to be careful to manage the pause/play state of `AudioQueue` with respect to both network conditions and the user's actions. If the user pauses during a network interruption, you want to stay paused when the data starts flowing again.

Dropped Connections

While Wi-Fi connections can be fairly reliable, you can count on cellular network connections disconnecting or timing out on a regular basis. For this reason, you may want to implement some form of connection healing that resumes dropped connections. Of course, you don't want to restart a download from the beginning of the file. So, the first thing you'll need is a server that supports delivery of a range of bytes from a file. Most modern web servers do this out of the box using the HTTP Range header.

Once you have the ability to download a partial file, the rest is straightforward code modification that can be limited to the `AudioRequest` class (this is one of the benefits of a clean code factoring). Let `AudioRequest` keep track of how many bytes have been downloaded, and when a connection drops, it can issue a new `NSURLConnection` request for the offset location where you left off. Figure 6-6 demonstrates how a file may be split into several requests due to dropped connections.

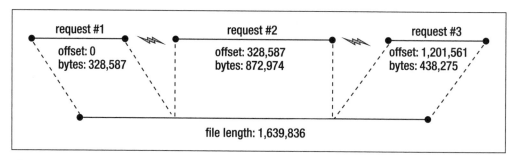

Figure 6-6. *Splitting a file into multiple requests*

The only remaining issue is the length of time-outs on `NSURLConnection`. This is a tricky issue: If your time-outs are too long, dropped connections will result in lengthy gaps in audio playback. If your time-outs are too aggressive, you risk making things worse for your servers when they are distressed.

For Pandora Radio, we've invested in a reliable server infrastructure, which allows us to have aggressive time-outs. Our (nonscientific) experimentation also shows that when cellular network connections become significantly delayed, they are unlikely to ever recover. For these reasons, our time-outs tend to be pretty aggressive. Your mileage may vary.

Minimizing Gaps Between Songs

Initiating a new network connections and loading enough data to begin playing a song can take a while, often 10 seconds or more on a cellular connection. When listening to a series of songs in a playlist, this can result in substantial gaps between songs, which is a less-than-stellar user experience.

To solve this problem, you can begin downloading audio for the upcoming song as the current song nears completion. You'll want to be careful, though, never to have two network connections open at the same time, or you effectively halve your bandwidth capacity for each song. Also beware: for processor-intensive audio encodings such as MP3 and AAC, the iPhone allows only one AudioQueue to exist at any time. So when preloading the next song, you must take care to make sure that next AudioQueue isn't created until after the current one is destroyed.

Resuming a Song

Since the iPhone is such an incredible, Swiss Army knife of a device, and since your application isn't allowed to run in the background, it's not surprising that your users will want to exit your application from time to time, if only briefly. Maybe a phone call or an important e-mail will come in. Whatever the reason, your user will feel comfort and ease if leaving your application won't result in missing the rest of an enjoyable song.

For this reason, we've invested much effort into supporting the ability to resume interrupted songs in Pandora Radio. This feature requires a three-pronged attack: knowing how many bytes have played in your song thus far, persisting that information when the application terminates (perhaps by using NSUserDefaults), and resuming the interrupted song on startup by initiating a partial download starting at the offset where you left off.

It's not necessary (and probably not a good idea) to resume songs that have been interrupted for extremely long periods of time. When you use an application after two days' absence, you may not remember what song you were listening to before. But if you leave the application to take a phone call and come back 10 minutes later, resuming interrupted songs is a great feature.

Improving Application Responsiveness

With all of the processing required to stream and play audio, it's inevitable that your audio-handling code may interfere with UI responsiveness in your application. There is no one-size-fits-all solution to this problem. In the case of Pandora Radio, we've worked to keep everything in the main run loop to maintain code simplicity, which means we do some work to make sure no one section of audio processing takes too long and holds up the run loop. Another option is to run audio code in a separate thread; the choice of solution is largely one of taste and preference.

No matter what your solution, always remember Jackson's first and second rules of optimization: "Don't do it." And "(for experts only) Don't do it yet."

Finding Help Resources

As you add features and make your audio more robust, you'll undoubtedly encounter problems for which you could use a little extra help. Here are some suggestions for where to start.

Core Audio documentation: If your application is the least bit interesting or successful, it won't be long before you'll have to dig deeper into the iPhone's audio technology to solve your problems. The Core Audio documentation is your first stop. Most of what you need is there, if you can piece it all together. The documentation can be found here:

```
http://developer.apple.com/iphone/library/documentation/MusicAudio/➥
Conceptual/CoreAudioOverview/Introduction/Introduction.html
```

Core Audio mailing list: The Core Audio group maintains a mailing list that is open to the public. Many audiophile programmers subscribe to and are active on the list. More importantly, Apple engineers read the list correspondence and respond on a regular basis. The mailing list is available at `http://lists.apple.com/mailman/coreaudio-api`. If you have a question, be sure to check the archives first. Core Audio newbies are always welcome, but with over eight years of archived history, you can bet that many questions have been asked and answered before.

Testing: Saving the Best for Last

I hope you've found our walkthrough of playing Internet audio on the iPhone informative and helpful so far. Before we part, I'd like to discuss the most fun part of building an audio application for iPhone—testing.

You've built your application, and it's functionally complete. Now what? For many applications, this moment induces a brief shudder as you realize it's time to test and debug. But for an audio application such as Pandora Radio, this shudder can instead be a jump for joy.

Don't get me wrong: there's no doubt you'll encounter many hard-to-debug issues once you start using your application. Debugging audio is hard. But finding and fixing bugs means lots of testing, and testing means you'll spend time listening to lots of great music.

And truly the best form of testing is to use your application as much as possible. Don't hold back—abuse it! While building Pandora Radio, we've taken our application for walks, bike rides, long car drives, elevator rides, and subway rides, just to name a few. We've also interrupted it with phone calls and SMS messages. We're constantly looking for conditions that can cause our application to misbehave, whether that's an application crash, a failure to recover from network loss, or even an excessively long audio delay. If we want to emulate the car radio, we need to be always on, which means all those difficult conditions need to be handled gracefully.

So *test your application*! It's fun, and it will improve your code quality immensely. Here are a couple audio URLs to kick off your testing: `http://neilmix.com/book/etude.mp3` and `http://neilmix.com/book/concerto.mp3`.

Happy bug hunting.

Summary

You now have all the tools you need to write audio on the iPhone: an understanding of how audio is transferred over a network, familiarity with how Core Audio translates audio data into sound on the device, and a preview of the challenges that lay ahead. Pandora Radio uses all the techniques discussed in this chapter, something I wouldn't have guessed was possible when we first started building the application. Even so, we're far from finished with Pandora Radio. I'm looking forward to further improvements in our own audio code, and more importantly, I'm eagerly anticipating innovations from other audio applications. It is my hope that you, as a potential audio developer on the iPhone, can take what's presented here and expand it in great, new, interesting ways.

Steven Peterson

Company: *Pixelcup*

Location: *San Francisco, California*

Former life as a developer: **One of the early developers of Yahoo's YUI user interface library, responsible for architecting several JavaScript components in YUI's first open source release. Also worked as a web applications developer using PHP, J2EE, and .NET for MetLife, Tommy Hilfiger, and IMVU.**

Life as an iPhone developer: **Routesy is featured in the App Store's Navigation category and was one of the first 500 applications to appear on the App Store on its original launch date. Routesy was painstakingly crafted using beta versions of Xcode bundled with the iPhone SDK prereleases.**

What's in this chapter: **This chapter discusses how to take an open XML data feed from the web and translate it into a hierarchical application that the user can navigate. It also covers how to take that data and put it into a location-sensitive context to help the user take advantage of the iPhone's geolocation capabilities.**

Key technologies

- **Core Location**
- **Accessing web services**
- **Table views and table view controllers**

Chapter 7

Going the Routesy Way with Core Location, XML, and SQLite

When I began writing Routesy, an application that allows San Francisco commuters to find out when the next bus will arrive, the dream of becoming a published iPhone software developer was the last thing on my mind. By the time Apple announced the iPhone SDK, I had already integrated the iPhone into my daily routine and was no longer hauling around an iPod and a separate phone in my pocket. The SDK convinced me that the iPhone is a device with unlimited potential, and I was determined to find a way to make my iPhone even more useful. Long before I ever even imagined that I might be selling downloadable software in the App Store, I was trying to figure out how to efficiently make my phone help me find the nearest bus and tell me when it would arrive.

My primary guiding principle while building Routesy was always no more complex than this mantra: "build something I'd want to use on a daily basis." Although I have real customers today, my first and most important goal was to build an application for myself. I recommend this approach to any new iPhone developer looking for something amazing to build. Solve a problem that you personally have, and do it well. Chances are that someone else has the same problem and will find your application useful.

In this chapter, we'll build a transit application from the ground up, using the iPhone's network and location-sensing technologies to put the transit predictions into a context where users will be able to find the nearest station and the next train arrival.

Starting from Scratch

Routesy was my first attempt at Cocoa development, so the learning curve was steep. I had to begin writing code in an unfamiliar programming language with a completely foreign tool set. While I'd like to say that I spent my first few weeks with the iPhone SDK building the foundation of the application I would eventually ship, the reality is quite different. I would describe it more as furiously clicking the Build button in Xcode after tweaking lines of code, just hoping to get my project to successfully compile.

There were two main stumbling blocks that I had to overcome through spending time experimenting with the SDK and the Apple-provided examples:

Objective-C: At first glance, the language looks far more complex than it actually is. The primary sticking point I encountered was the distinctive bracket-based syntax. In other C-based languages, such as Java, you might be used to calling a method by invoking a call using a dot notation such as `myObject.doSomething()`. Objective-C uses brackets to group together the calling object and the method: `[myObject doSomething];`. It may take you a while to get used to seeing so many brackets in your code, but rest assured that this notation actually makes it easier to read long, nested method calls. You don't have to master Objective-C before building your iPhone application, and you'll probably find that it's easy to learn as you go.

Interface Builder: I initially found the concept of connecting outlets and actions in Interface Builder to the objects in my application to be a bit daunting. I found that the most helpful resources for understanding Interface Builder were the project templates included with Xcode. By providing several starter templates that you can use as the basis for your project, Apple gives you example applications with Interface Builder XIB files that already have connections set up in your code. The Apple example projects included with the SDK are also great resources for understanding how to effectively separate user interfaces from your code using Interface Builder.

Assessing the Application Requirements

Routesy currently supports the San Francisco MUNI subway and bus lines. For this exercise, we're going to build a version of Routesy that will allow users to view real-time predictions for BART (Bay Area Rapid Transit), another popular public transit provider in the Bay Area. Let's get started by assessing the requirements for our new application.

- *It needs to be fast!* This is the most important point. BART already has a mobile web site for checking train arrival predictions, so why would people want to use a native iPhone application? The biggest benefit of using a native application is speed. We can cache static data, such as the list of stops, so that the user isn't burdened with downloading unnecessary data every time the application is used.

- *It needs location awareness*: The application should take advantage of the iPhone's location-sensing capabilities to make finding the nearest stop easy and to allow the user to get predictions without having to scroll through a long list of train stations.

- *It needs simple navigation*: The application should launch displaying a table view of stops, sorted by the nearest one. When the user taps a stop, the application should display a list of real-time predictions for the selected stop, retrieved from the BART XML feed, available at `http://www.bart.gov/dev/eta/bart_eta.xml`.

The first step for building our application is to create the static database that contains the list of stations. First, though, we should examine the data feed provided by BART to get a better idea of what data we'll need to store.

The BART XML feed, located at `http://www.bart.gov/dev/eta/bart_eta.xml`, consists of multiple `station` elements, each of which contains information about the station and a list of destinations and arrival estimates:

```
<station>
  <name>12th St. Oakland City Center</name>
  <abbr>12TH</abbr>
  <date>02/10/2009</date>
  <time>10:29:00 PM PST</time>
  <eta>
    <destination>Fremont</destination>
    <estimate>4 min, 17 min</estimate>
  </eta>
</station>
```

The XML feed provided by BART is useful for loading train prediction estimates, but it doesn't contain the latitude and longitude data you will need to sort stops by the user's current location. We can use the station name from the `name` element and the unique identifier from the `abbr` element in our database, but we will also need the latitude and longitude for each station.

I retrieved this data using a simple screen-scraping command-line application that loads the station page for each BART station from BART's web site and extracts the coordinates from the map image displayed on the page. The scraper application is beyond the scope of this exercise, but the final database file containing all 43 records is included with the sample project, and I've also included the scraper application (*RoutesyBARTDB.xcodeproj*) so you can see how the database was built.

TIP

> You may find it easier to debug your application using a SQLite administration application to browse your database and test SQL queries. While developing Routesy, I used SQLite Database Manager (`http://sqlitebrowser.sourceforge.net`), an open source freeware utility.

Creating the Routesy User Interface and Classes

The iPhone SDK includes several handy templates for building your iPhone applications without having to go through a lot of mundane setup tasks each time you create a new project. In this application, we're going to set up two screens, each with a `UITableView`. Apple provides a class called `UINavigationController` that allows you to set up a view stack for showing data in a hierarchy, which is exactly what we need for this application. Thankfully, the SDK has a template for a navigation-based application, which is where we will start our project.

1. In Xcode, create a new project by choosing **File ➤ New Project**, and choose Navigation-Based Application, as shown in Figure 7-1.

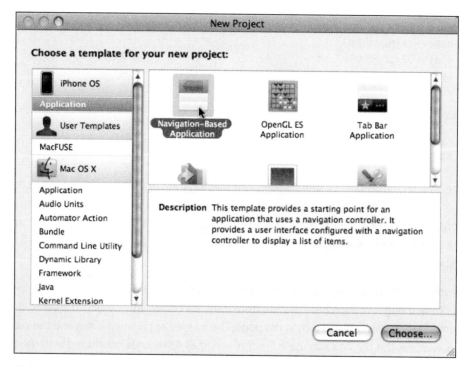

Figure 7-1. Choosing Navigation-Based Application from the New Project dialog

2. Name your project RoutesyBART. After you've created the project, you'll be presented with a list of automatically generated groups and files, as shown in Figure 7-2.

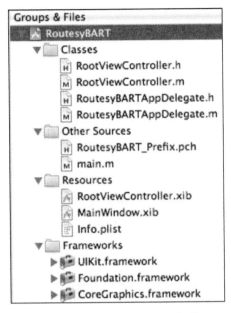

Figure 7-2. *Files that are automatically generated after choosing a template*

Let's start by taking an inventory of what files were automatically generated or included for free with your application when you created your project:

- Frameworks

  - *UIKit.framework*: The primary framework used by the iPhone to build and display user interface elements.

  - *Foundation.framework*: The most basic framework used by all iPhone applications. This framework contains frequently used classes like NSString, NSArray, and NSNumber.

  - *CoreGraphics.framework*: A C-based framework that handles 2D drawing using the Quartz drawing engine. Many of the UIKit classes use Core Graphics to draw their user interface elements.

- Classes

  - *RootViewController.m*: The main table view controller that will contain the initial list of BART stations from which the user will make a selection.

- *RoutesyBARTAppDelegate.m*: The delegate class that contains the high-level code that adds your navigation controller to the application window after the application launches.

- *main.m*: You should never need to modify this file. It contains the main application run loop and is set up for you.

- User interface files

  - *MainWindow.xib*: This Interface Builder XIB file contains an instance of a navigation controller that will allow your application to manage a stack of views. It also contains a reference to your application delegate and window object.

  - *RootViewController.xib*: Another XIB file that contains the UITableView that you will use to display the list of BART stations to the user when the application first opens.

- Other files

  - *Info.plist*: A list of properties about your application like name and version number. We'll revisit this file later when we're putting the finishing touches on the application.

Before we continue, you should set your project to run using the iPhone Simulator included with the SDK. You can test your application on your phone later, but it will be faster to debug by initially using the simulator.

3. Select the iPhone Simulator for the latest installed SDK version by choosing the simulator under Active SDK in the drop-down at the top of the Xcode window, as shown in Figure 7-3.

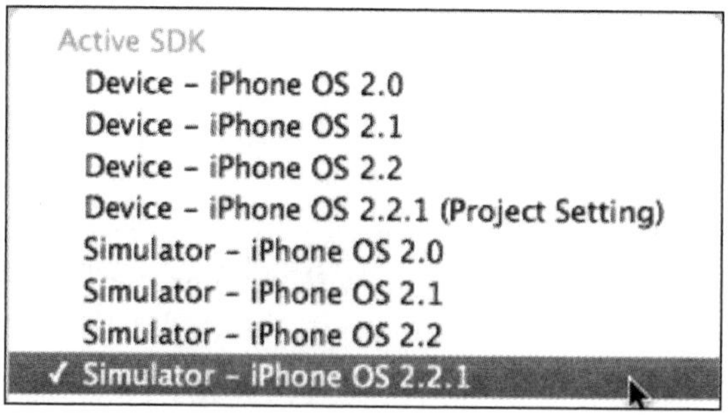

```
Active SDK
    Device – iPhone OS 2.0
    Device – iPhone OS 2.1
    Device – iPhone OS 2.2
    Device – iPhone OS 2.2.1 (Project Setting)
    Simulator – iPhone OS 2.0
    Simulator – iPhone OS 2.1
    Simulator – iPhone OS 2.2
 ✓  Simulator – iPhone OS 2.2.1
```

Figure 7-3. *Choosing the iPhone Simulator from the Active SDK drop-down*

4. Add the static database (*routesy.db*) to the Resources folder of your project by select-ing the Resources folder, selecting **Project ➤ Add to Project**, and selecting the database file location in the project examples folder. After you select the database file, you'll be presented with a dialog like the one shown in Figure 7-4. Make sure to leave "Copy items into destination group's folder" unchecked, and click the Add but-ton to add the database to your project.

Figure 7-4. *Adding the static database file to the project*

If you update your database in the future, the new version will automatically be cop-ied when you build your project, since your project will only contain a reference to the database, rather than a copy of the file.

5. Next, you'll need to add references to a few more frameworks and dynamic libraries that your application will use. To add references to the libraries, choose **Project ➤ Edit Active Target "RoutesyBART"**, and in the General tab, you'll see a list of linked libraries at the bottom of the panel. Click the + button, and click *CoreLocation. framework*, *SystemConfiguration.framework*, *libsqlite3.dylib*, and *libxml2.dylib*. You can select all four libraries at the same time by command-clicking each and clicking the Add button, as shown in Figure 7-5.

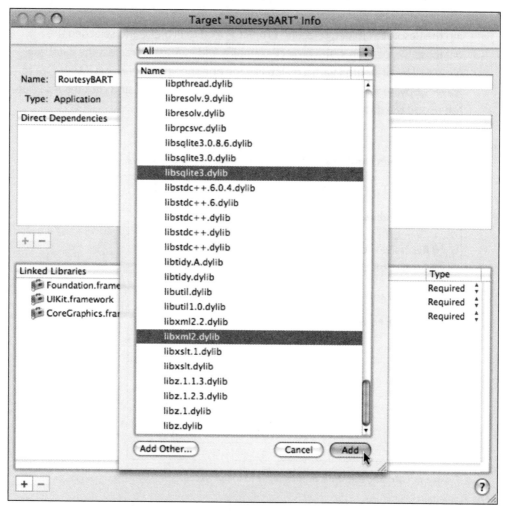

Figure 7-5. *Selecting frameworks and libraries to link to the project*

Let's take a moment to become familiar with the libraries and frameworks that we just linked to the project:

- *CoreLocation.framework*: This framework provides us with access to the iPhone's location API, which your application uses to help the user find the nearest BART station.

- *SystemConfiguration.framework*: This contains APIs that allow us to determine the configuration of the user's device. In the case of Routesy, we need to make sure the network is available before attempting to retrieve prediction data.

- *libsqlite3.dylib*: This dynamic C library provides an API for querying the static SQL database included with our project.

- *libxml2.dylib*: This dynamic library gives the application access to fast parsing of XML documents and support for XPath querying, which will help us quickly find the prediction data the user requests.

6. The libxml2 library also requires that you include a reference to the libxml header files, which are located on your system in the path */usr/include/libxml2*. To add the headers, select the Build tab in the Target Info window that we've already opened, and add the path to the Header Search Paths field, as shown in Figure 7-6.

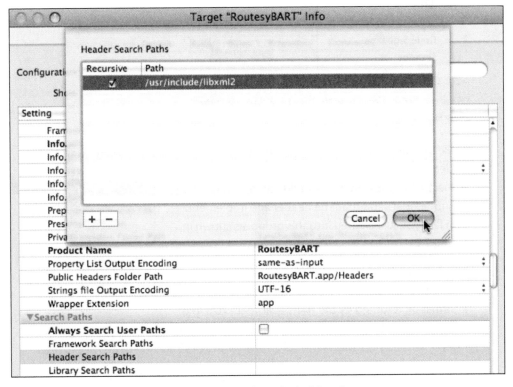

Figure 7-6. *Adding the libxml2 header search paths to the build settings*

Invoke the dialog shown in Figure 7-6 by double-clicking the Header Search Paths field. Make sure that you check the Recursive box so that your application can find all of the libxml2 headers that you'll need to build your application.

Now that all of your project dependencies have been set up, we're ready to get started coding. First, we need to build the model for your project—the objects that will represent pieces of data. For Routesy, there are two types of objects: a station and a prediction. You'll start by creating an object for each.

7. Choose **File ➤ New File**, and under Cocoa Touch Classes, select "NSObject subclass," as shown in Figure 7-7. One of the files will be called *Station.m*, and the other will be called *Prediction.m*. When you create these classes, each implementation file (with extension .*m*) will automatically be accompanied by a header file (with extension .*h*).

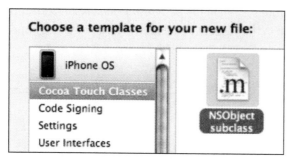

Figure 7-7. *Creating a new NSObject subclass file in the New File dialog*

8. To keep your files organized, let's also create a folder (referred to in Xcode as a **group**) by choosing **Project ➤ New Group**. You should name your new group "model", and drag the header and implementation files you just created into this group, as shown in Figure 7-8.

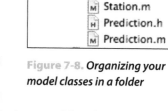

Figure 7-8. *Organizing your model classes in a folder*

The structure for these classes is very basic, and maps closely to the data in our static database and the information returned by the BART XML feed. To avoid memory leaks, don't forget to release instance variables in your objects' deal1oc methods. Listing 7-1 shows the code for your Station and Prediction classes.

Listing 7-1. *Creating the Station and Prediction Model Classes*

```
//
//  Station.h
//

#import <Foundation/Foundation.h>

@interface Station : NSObject {
  NSString *stationId;
  NSString *name;
  float latitude;
  float longitude;
  float distance;
}
```

```objc
@property (copy) NSString *stationId;
@property (copy) NSString *name;
@property float latitude;
@property float longitude;
@property float distance;
@end


//
//  Station.m
//

#import "Station.h"

@implementation Station
@synthesize stationId, name, latitude, longitude, distance;

- (void)dealloc {
  [stationId release];
  [name release];
  [super dealloc];
}

@end


//
//  Prediction.h
//

#import <Foundation/Foundation.h>

@interface Prediction : NSObject {
  NSString *destination;
  NSString *estimate;
}

@property (copy) NSString *destination;
@property (copy) NSString *estimate;

@end


//
//  Prediction.m
//

#import "Prediction.h"

@implementation Prediction
@synthesize destination, estimate;
```

```
- (void)dealloc {
  [destination release];
  [estimate release];
 [super dealloc];
}
```

@end

Next, we'll deal with the controllers in your project. Roughly speaking, a **controller** is an object that bridges your application's model (the objects that contain data) with the **view** (what the application displays to the user).

There is already a controller in your project, *RootViewController.m*, which is the class for the initial UITableViewController that is displayed when the user launches your application. We'll need a second table view controller to manage the list of predictions the user will see when selecting a station, so let's create a class for that too.

9. Choose **File ➤ New File**, and this time choose "UITableViewController subclass" as your template, as shown in Figure 7-9. Call your new class PredictionTableViewController.

To keep things organized, now would be a good time to create a group called "controller" in which to keep your controller classes, just like you did for your model classes in Figure 7-8. You should place both RootViewController and PredictionTableViewController in this new group.

Figure 7-9. Creating a UITableViewController subclass

Both of these view controller classes have a ton of handy, commented method implementations in place to help us remember what we need to implement to get our table views up and running. We'll implement some of these methods later as we begin to add to our application's functionality.

At this point, we have a great starting point to begin showing the list of stations in the initial table view.

10. First, we need to add a property to RootViewController so we have somewhere to store the list of station objects. Add an instance variable to *RootViewController.h*:

NSMutableArray *stations;

11. Also, add a property to the header:

@property (nonatomic,retain) NSMutableArray *stations;

12. At the top of *RootViewController.m*, in the implementation section, make sure to synthesize your new property:

```
@synthesize stations;
```

13. Now, you need to open the database, retrieve the list of stations, and put that list into the mutable array that we just created. We only need to load the static list of stations once when the application starts since the list is unchanging, so we'll load the list by implementing the `viewDidLoad` method of `RootViewController`.

    The code in Listing 7-2 initializes the `stations` array and executes a query against the database file to get the list of stations. For each row in the database, you'll allocate a new `Station` object and add it to the array, as shown in Listing 7-2. You'll notice that this code makes extensive use of SQLite C APIs, which you can read about in more detail at http://www.sqlite.org, or in *The Definitive Guide to SQLite* by Mike Owens (Apress, 2006).

Listing 7-2. *Loading the Station List from the Database*

```
- (void)viewDidLoad {
    [super viewDidLoad];

    // Load the list of stations from the static database
    self.stations = [NSMutableArray array];

    sqlite3 *database;
    sqlite3_stmt *statement;

    NSString *dbPath = [[NSBundle mainBundle]
                        pathForResource:@"routesy" ofType:@"db"];

    if (sqlite3_open([dbPath UTF8String], &database) == SQLITE_OK) {
        char *sql = "SELECT id, name, lat, lon FROM stations";
        if (sqlite3_prepare_v2(database, sql, -1, &statement, NULL)
            == SQLITE_OK) {

            // Step through each row in the result set
            while (sqlite3_step(statement) == SQLITE_ROW) {
                const char* station_id =
                    (const char*)sqlite3_column_text(statement, 0);
                const char* station_name =
                    (const char*)sqlite3_column_text(statement, 1);
                double lat = sqlite3_column_double(statement, 2);
                double lon = sqlite3_column_double(statement, 3);
```

```
            Station *station = [[Station alloc] init];
            station.stationId = [NSString stringWithUTF8String:station_id];
            station.name = [NSString stringWithUTF8String:station_name];
            station.latitude = lat;
            station.longitude = lon;

            [self.stations addObject:station];
            [station release];
        }
        sqlite3_finalize(statement);
    }
    sqlite3_close(database);
  }
}
```

14. To get the `UITableView` to display rows, you need to implement three basic methods. First, you need to set the number of sections that your table view has—in this case, one. This function is already implemented in the template for *RootViewController.m*:

```
- (NSInteger)numberOfSectionsInTableView:(UITableView *)tableView {
    return 1;
}
```

15. Next, the `UITableView` needs to know how many table cells to display. This is as simple as returning the number of rows in the array of stations:

```
- (NSInteger)tableView:(UITableView *)tableView
        numberOfRowsInSection:(NSInteger)section {
    return [self.stations count];
}
```

16. Finally, we'll implement the method that determines what value to display in a cell when the table view is displayed.

The iPhone SDK uses a clever method of keeping memory usage at a manageable level when scrolling through long lists of items in a table view. Instead of creating a cell for each item, which could use vast amounts of memory, the table only allocates as many cells as can be displayed at once, and when a cell scrolls out of the viewable area, it is queued up to be reused when a new cell needs to be displayed.

The following method always checks to see if there is an already allocated cell available to be reused by calling dequeueReusableCellWithIdentifier each time a cell is displayed. The CellIdentifier string allows your table view to have more than one type of cell. In this case, we'll set the identifier to "station".

To determine which station corresponds with the cell being displayed, this method provides a handy NSIndexPath object, which has a property called row. You'll see from the code below that we use the row index to retrieve a Station object from the stations array, and once we have a cell to work with, we can set the text property of the cell to the name of the station, as shown in Listing 7-3.

Listing 7-3. *Setting Up the Station List Table Cell Text*

```
- (UITableViewCell *)tableView:(UITableView *)tableView
      cellForRowAtIndexPath:(NSIndexPath *)indexPath {

    static NSString *CellIdentifier = @"station";
    Station *station = [self.stations objectAtIndex:indexPath.row];

    UITableViewCell *cell = [tableView
        dequeueReusableCellWithIdentifier:CellIdentifier];
    if (cell == nil) {
        cell = [[[UITableViewCell alloc] initWithFrame:CGRectZero
            reuseIdentifier:CellIdentifier] autorelease];
    }

    // Set up the cell...
    cell.text = station.name;
    return cell;
}
```

17. At the top of *RootViewController.m*, you'll need to add two additional #import state-ments to include dependencies that your new code relies on. At the top of the file, add the following lines so your project will compile properly:

```
#import <sqlite3.h>
#import "Station.h"
```

With the table view code in place, we can finally test Routesy for the first time. In Xcode, click "Build and Go" in the toolbar, and your application will compile and launch in the iPhone Simulator. Once the application launches, you'll be presented with a view like the one shown in Figure 7-10.

There really isn't much to see yet. You'll be able to scroll through the list of stops that are being loaded from your database, but selecting a row won't do anything yet.

The next step will be to properly set up the user interface so that tapping a station name will allow the user to see a list of predictions for that station.

Figure 7-10. *Your station list UITableView in action*

We already have a class for the table view controller that will display predictions: PredictionTableViewController. However, up to this point, we haven't created an instance of this class to display.

You may have already noticed that nowhere in the code do we create any instances of RootViewController either. This is because the project template uses Interface Builder to create an instance of RootViewController for us. You will mirror this approach when creating an instance of PredictionTableViewController.

Make sure to save any unsaved files in your project, and then under the Resources folder in your project, double-click *MainWindow.xib* to open the user interface file in Interface Builder. Two windows will be displayed: the document, shown in Figure 7-11, and the window for the navigation controller that the application uses to navigate back and forth and to manage the stack of visible views, shown in Figure 7-12.

Figure 7-11. *The default Interface Builder document view*

Figure 7-12. *The Navigation Controller window in Interface Builder*

Many of the objects we'll be working with are in the hierarchy of views under Navigation Controller, and these views are impossible to see unless you change the document View Mode to something more friendly. For this project, you'll use the List View, which you can enable by clicking the center button above View Mode in the document toolbar, shown in Figure 7-13.

Figure 7-13.
Changing the View Mode to List in Interface Builder

Now is a good time to become familiar with the way that the project template has set up this primary user interface file. The navigation controller provided for us by default has a navigation bar and an instance of RootViewController, the top-level class that was automatically generated for you when you created the project and that currently contains the table view that displays the list of stations.

18. Underneath Root View Controller is an instance of UINavigationItem, which contains information about how this view controller fits into our navigation hierarchy. Let's see the information for our root controller by expanding the controller and clicking the navigation item. Then, in Interface Builder, choose **Tools ➤ Inspector**. When the Inspector window pops up, choose the first tab, called Attributes, as shown in Figure 7-14.

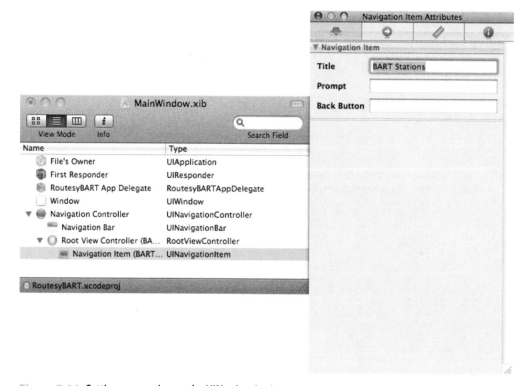

Figure 7-14. *Setting properties on the UINavigationItem*

The navigation item has three text attributes displayed in the Inspector window: Title, Prompt, and Back Button. Title determines what should be displayed in the bar at the top of the screen when the navigation item's view controller is visible in the view stack. The Prompt field allows your application to display a single smaller line of text above the title, and the Back Button field contains the text that will be displayed on the next screen inside the back button that will take the user back to this screen.

19. For now, let's simply type **BART Stations** into the Title field.

Next, we're ready to create an instance of `PredictionTableViewController`, the class you created earlier in step 9, to use for the second screen.

20. Open the Library window by clicking **Tools ➤ Library**. Under Cocoa Touch Plugin, in the Controllers section, grab an instance of Table View Controller, and drag it to the bottom of your document, as shown in Figure 7-15.

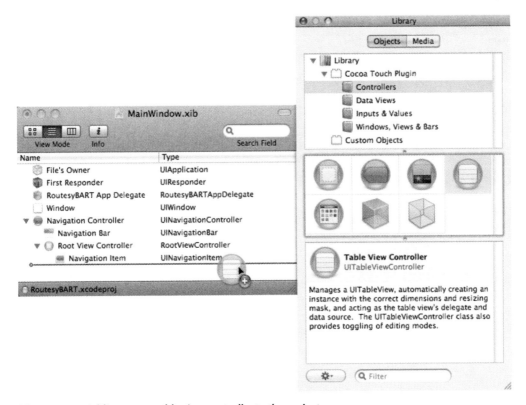

Figure 7-15. *Adding a new table view controller to the project*

Initially, the class for our new table view controller is set to `UITableViewController`. However, since we already created our own custom controller class, `PredictionTableViewController`, we need to tell Interface Builder that the new controller's type should match the class we created.

21. Select the table view controller you just dragged into the document, and open the Inspector window again. Choose the Identity tab, and in the Class field, set the class to `PredictionTableViewController`, as shown in Figure 7-16.

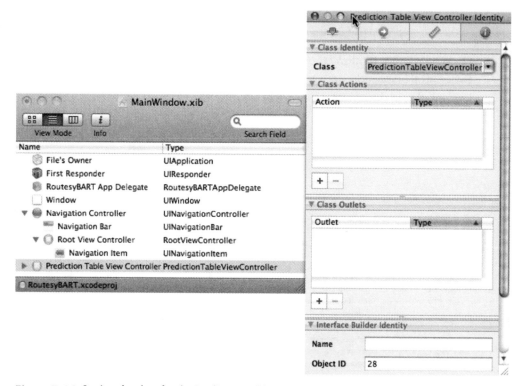

Figure 7-16. *Setting the class for the Prediction Table View Controller*

22. Our second table view controller will also need a navigation item so that we can set a title for the predictions table. From the Library, drag a navigation item onto your new table view controller, and set the title in the Inspector window the same way you did for the root view controller object; see Figure 7-17. Be sure to drop this new navigation item inside the Prediction Table View Controller as shown so that it's associated with the proper view controller.

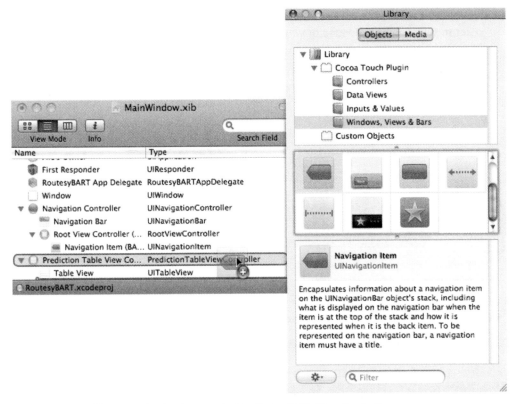

Figure 7-17. *Adding a navigation item to the Prediction Table View Controller*

Now, you have an instance of PredictionTableViewController loaded into your user interface XIB, ready to be used in your application. Save the changes in Interface Builder and return to Xcode.

23. Next, we need to enable the table on the first screen to push the new second table onto the view stack when the user selects a station. Inside the interface for RootViewController in *RootViewController.h*, add #import "PredictionTableViewController.h" to your #import statements, and then declare a new property that you will use to reference your new controller instance:

```
PredictionTableViewController *predictionController;
```

24. Set up the property for this new instance variable, but this time, add a reference to IBOutlet in front of the type in the property declaration. This tells Interface Builder that you want the ability to connect an object in Interface Builder with this property. Don't forget to also synthesize the property in *RootViewController.m*.

```
@property (nonatomic,retain)
    IBOutlet PredictionTableViewController *predictionController;
```

25. Save your changes, switch back to Interface Builder, and click the Root View Controller to select it. In the Inspector window, choose the second tab, called Connections.
You'll see that there is now an outlet to your `predictionController` instance. Connect that outlet to your Prediction Table View Controller by dragging the circle from
the outlet to the controller in the document window, as shown in Figure 7-18.

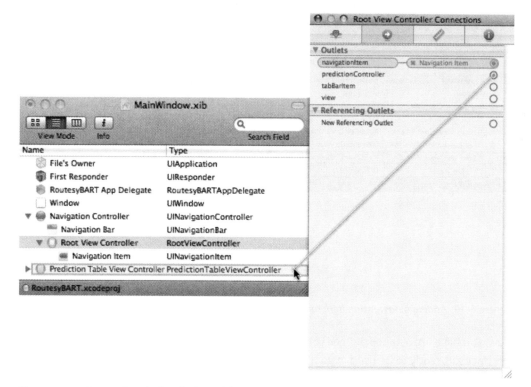

Figure 7-18. *Connecting the Prediction Table View Controller to the predictionController outlet*

26. Now that your root view controller has access to the new prediction controller that
we created, you can set up the root controller to reveal the predictions table when
you tap on a station name. Table view controllers have a delegate method called
`didSelectRowAtIndexPath` that you can implement that will be called whenever
you select an item in the table view. We will implement this method to tell our
application to push the prediction controller onto the view stack when you tap a
selection:

```
- (void)tableView:(UITableView*)tableView
              didSelectRowAtIndexPath:(NSIndexPath*)indexPath {
    [self.navigationController
          pushViewController:self.predictionController animated:YES];
}
```

27. We also need a way to tell the prediction controller what station the user selected. We can do that by grabbing a reference to the station object and setting it into a property on the prediction controller. Create an instance variable and property called `station` on the prediction controller for the station. The type for this property is `Station`, and you should create and synthesize this property on `PredictionTableViewController` in the same way you've created properties for other classes during this exercise. When creating new properties and instance variables, don't forget to release them in your class's `dealloc` method as well, to avoid memory leaks.

28. Then, back in `RouteViewController`, we can get a reference to the selected station using the row index provided in `indexPath` and set it into the new property we just created, so that the prediction controller knows which station was selected.

```
- (void)tableView:(UITableView*)tableView
didSelectRowAtIndexPath:(NSIndexPath*)indexPath{
        Station *selectedStation =
                    [self.stations objectAtIndex:indexPath.row];
        self.predictionController.station = selectedStation;
        [self.navigationController
            pushViewController:self.predictionController animated:YES];
}
```

`UIViewController`, the prediction controller's base class, has a method called `viewWillAppear` that is called before a view is displayed to the user. We can implement this method on `PredictionTableViewController` to set the title that will be displayed on the screen before the prediction screen is displayed to the user.

29. In *PredictionTableViewController.m*, implement this method:

```
- (void)viewWillAppear:(BOOL)animated {
    [super viewWillAppear:animated];
    self.title = self.station.name;
}
```

30. Build and run your application again, and when you tap a station name, you'll see your new table view controller slide into view with the name of the station you selected displayed at the top of the second screen. We don't have any data to display yet, so that will be the next big step.

Bringing Real-Time Predictions to Routesy

Now that we have your model and controllers in place, we're ready to start loading real-time predictions from the BART data feed. To keep our application nice and clean, we'll encapsulate the logic for loading feed data into a new class, called `BARTPredictionLoader`.

1. Create a new class by choosing **File ➤ New File**, and choose NSObject as the base class for your new class.

There are a few things we'll need to make our BARTPredictionLoader class as useful as possible. We're going to create a method that will asynchronously grab the XML data from the BART feed. We'll also create a custom protocol so that we can assign a callback delegate, so that our code can be easily notified when the XML data has finished loading. There will be two NSMutableData properties: one for the data we're loading and a copy of the last successfully loaded data. Finally, we'll make a singleton instance of BARTPredictionLoader that your application can access from anywhere. Listing 7-4 shows what your header definition should look like.

Listing 7-4. *Creating the BARTPredictionLoader Interface*

```
//
//   BARTPredictionLoader.h
//

#import <Foundation/Foundation.h>
#import <SystemConfiguration/SystemConfiguration.h>

@protocol BARTPredictionLoaderDelegate
- (void)xmlDidFinishLoading;
@end

@interface BARTPredictionLoader : NSObject {
    id _delegate;
    NSMutableData *predictionXMLData;
    NSMutableData *lastLoadedPredictionXMLData;
}

+ (BARTPredictionLoader*)sharedBARTPredictionLoader;
- (void)loadPredictionXML:(id<BARTPredictionLoaderDelegate>)delegate;

@property (nonatomic,retain) NSMutableData *predictionXMLData;
@property (nonatomic,retain) NSMutableData *lastLoadedPredictionXMLData;

@end
```

2. Next, you'll need to actually implement the code to load data from the BART feed. Let's start by implementing loadPredictionXML. Notice that this method takes as an argument a delegate object that implements our protocol, BARTPredictionLoaderDelegate. Our code will set the delegate into the _delegate instance variable, where we'll keep it until we need it.

3. Before attempting to call the network, you should make sure that the network is currently available on the iPhone. The SCNetworkReachability functions provided by *SystemConfiguration.framework* will allow you to do just that.

4. Assuming that the reachability flags indicate that the network is available, you can use NSURLConnection to create an asynchronous connection to load the data from the BART feed, as shown in Listing 7-5.

Listing 7-5. *Checking the Network and Creating a Connection*

```
- (void)loadPredictionXML:(id<BARTPredictionLoaderDelegate>)delegate {
  _delegate = delegate;

  // Load the predictions XML from BART's web site
  // Make sure that bart.gov is reachable using the current connection

  SCNetworkReachabilityFlags  flags;
  SCNetworkReachabilityRef reachability =
    SCNetworkReachabilityCreateWithName(NULL,
                                        [@"www.bart.gov" UTF8String]);
  SCNetworkReachabilityGetFlags(reachability, &flags);

  // The reachability flags are a bitwise set of flags
  // that contain the information about
  // connection availability
  BOOL reachable = ! (flags &
                      kSCNetworkReachabilityFlagsConnectionRequired);

  NSURLConnection *conn;
  NSURLRequest *request = [NSURLRequest
    requestWithURL:[NSURL
      URLWithString:@"http://www.bart.gov/dev/eta/bart_eta.xml"]];
  if ([NSURLConnection canHandleRequest:request] && reachable) {
    conn = [NSURLConnection connectionWithRequest:request delegate:self];
    if (conn) {
        self.predictionXMLData = [NSMutableData data];
    }
  }
}
```

5. NSURLConnection's connectionWithRequest method also takes a delegate argument. In this case, we'll set the delegate to self, so that we can implement the connection's delegate methods right here in the BARTPredictionLoader class. NSURLConnection has several delegate methods, three of which we'll implement: didReceiveResponse, didReceiveData, and connectionDidFinishLoading. The comments in Listing 7-6 explain how each of the delegate methods works, while Figure 7-19 shows the order in which these delegate methods are called.

Listing 7-6. *The NSURLConnection's Delegate didReceiveResponse Method*

```
- (void)connection:(NSURLConnection *)connection
            didReceiveResponse:(NSURLResponse*)response {
    // didReceiveResponse is called at the beginning of the request when
    // the connection is ready to receive data. We set the length to zero
    // to prepare the array to receive data
    [self.predictionXMLData setLength:0];
}


- (void)connection:(NSURLConnection *)connection
    didReceiveData:(NSData *)data {
    // Each time we receive a chunk of data, we'll appeend it to the
    // data array.
    [self.predictionXMLData appendData:data];
}


- (void)connectionDidFinishLoading:(NSURLConnection *)connection {
    // When the data has all finished loading, we set a copy of the
    // loaded data for us to access. This will allow us to not worry about
    // whether a load is already in progress when accessing the data.

    self.lastLoadedPredictionXMLData = [self.predictionXMLData copy];

    // Make sure the _delegate object actually has the xmlDidFinishLoading
    // method, and if it does, call it to notify the delegate that the
    // data has finished loading.
    if ([_delegate respondsToSelector:@selector(xmlDidFinishLoading)]) {
      [_delegate xmlDidFinishLoading];
    }
}
```

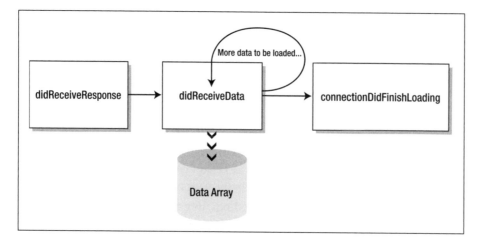

Figure 7-19. *NSURLConnection's delegate methods*

6. Finally, you'll need to set up a singleton instance of BARTPredictionLoader so that you can easily call it from anywhere in your application's code (see Listing 7-7). The @synchronized block around the initialization of the prediction loader ensures that the instance you create will be thread-safe. Refer to Chapter 3 for more detailed information on how threading works.

With this class method in place, you'll be able to access the shared instance of the prediction loader from anywhere by simply calling [BARTPredictionLoader sharedBARTPredictionLoader]. For a full explanation of how to properly implement singletons, see Apple's *Cocoa Fundamentals Guide*:

```
http://developer.apple.com/documentation/Cocoa/Conceptual/
CocoaFundamentals/CocoaObjects/CocoaObjects.html#//apple_ref/doc/uid
/TP40002974-CH4-SW32
```

Listing 7-7. *Creating a Shared Instance of BARTPredictionLoader*

```
static BARTPredictionLoader *predictionLoader;

+ (BARTPredictionLoader*)sharedBARTPredictionLoader {

    @synchronized(self) {
      if (predictionLoader == nil) {
        [[self alloc] init];
      }
    }
    return predictionLoader;
}
```

Now that we have the ability to load our data on demand, we need to figure out where to use it. Let's look again at the BART feed at http://www.bart.gov/dev/eta/bart_eta.xml.

You'll notice that the feed contains predictions for all the stations in one very small, quick-to-load XML file. Normally, we would only load predictions for the station that a user selects, but this file contains everything, so it makes the most sense to begin loading the data when our list of stations first loads, so we can be ready with predictions when the user chooses a station.

7. Let's revisit the viewDidLoad code in *RootViewController.m*. First, we need to keep the user from selecting anything in the table until the predictions are done loading. Then, we'll begin loading the XML. Add #import "BARTPredictionLoader.h" to the imports in *RootViewController.m*, and then add the following code to the end of viewDidLoad:

```
self.tableView.userInteractionEnabled = NO;
[[BARTPredictionLoader sharedBARTPredictionLoader] loadPredictionXML:self];
```

8. Remember when we created a protocol for the prediction loader? We need to attach this protocol to `RootViewController` to tell `BARTPredictionLoader` that `RootViewController` wants to be notified when the data finishes loading. In *RootViewController.h*, you can add the protocol to the end of the interface declaration:

```
@interface RootViewController :
         UITableViewController <BARTPredictionLoaderDelegate> { ...
```

9. Now, we can implement the protocol's `xmlDidFinishLoading` method in `RootViewController` so that we can reenable the table after the XML loads.

```
- (void)xmlDidFinishLoading {
   self.tableView.userInteractionEnabled = YES;
}
```

With that out of the way, we can now focus on loading the correct predictions for the selected station, which means that we need a way to query the loaded XML to get the predictions for the selected stop. We're going to query the XML loaded by `BARTPredictionLoader` using the XPath implementation provided by libxml2, which we included when we initially created the project.

Matt Gallagher, author of the popular Cocoa With Love blog (`http://cocoawithlove.com`), provides for free use a set of wrapper functions for performing XPath queries. Since libxml2's C API can be difficult to work with, Matt's `PerformXMLXPathQuery` function will save us lots of extra time and effort.

10. Now, we'll add a method to `BARTPredictionLoader` called `predictionsForStation` that takes the unique station ID as an argument, as shown in Listing 7-8. We'll use this XPath query to find the `eta` elements that match the unique station ID: `//station[abbr='%@']/eta`. The `PerformXMLXPathQuery` function returns an array of dictionaries containing estimates and destinations for the station.

TIP

Apple's *Event-Driven XML Programming Guide for Cocoa* (`http://developer.apple.com/iphone/library/documentation/Cocoa/Conceptual/XMLParsing/XMLParsing.html`) lists several helpful resources for working with XML in Cocoa applications.

Listing 7-8. *Loading the Real-Time Predictions for a Station*

```
- (NSArray*)predictionsForStation:(NSString*)stationId {
   NSMutableArray *predictions = nil;
```

```
  if (self.predictionXMLData) {
    NSString *xPathQuery = [NSString stringWithFormat:
                              @"//station[abbr='%@']/eta", stationId];
    NSArray *nodes =
        PerformXMLXPathQuery(self.predictionXMLData, xPathQuery);
    predictions = [NSMutableArray arrayWithCapacity:[nodes count]];

    NSDictionary *node;
    NSDictionary *childNode;
    NSArray *children;

    Prediction *prediction;
    for (node in nodes) {
      children = (NSArray*)[node objectForKey:@"nodeChildArray"];
      prediction = [[Prediction alloc] init];
      for (childNode in children) {
          [prediction setValue:[childNode objectForKey:@"nodeContent"]
                      forKey:[childNode objectForKey:@"nodeName"]];
      }
      if (prediction.destination && prediction.estimate) {
        [predictions addObject:prediction];
      }
      [prediction release];
    }
    NSLog(@"Predictions for %@: %@", stationId, predictions);
    }
  return predictions;
}
```

11. The `PredictionTableViewController` class needs a property called `predictions` to hold the list of predictions that the table will display. Before continuing, you should declare a property on `PredictionTableViewController` of type NSArray, similar to the one you declared on RootViewController, `stations`.

With this new property in place, we can implement the `PredictionTableViewController` viewWillAppear method, which will set the predictions into the prediction controller before the view appears. We also need to reload the table data each time the view appears since the user may go back and change the active station. Our viewWillAppear method now will look like Listing 7-9.

Listing 7-9. *Loading the Predictions Before the View Appears*

```
- (void)viewWillAppear:(BOOL)animated {
    [super viewWillAppear:animated];
    self.title = self.station.name;
```

```
    self.predictions = [[BARTPredictionLoader sharedBARTPredictionLoader]
                            predictionsForStation:self.station.stationId];
    [self.tableView reloadData];
}
```

12. Finally, we're ready to start displaying prediction data. Specifically, our table cells will display the estimate value for each Prediction in the predictions array. You should implement the three table view methods in PredictionTableViewController the same way you did for RootTableViewController. As a reminder, you'll need to implement numberOfSectionsInTableView, numberOfRowsInSection, and cellForRowAtIndexPath. Once you're got those methods in place, you're ready to see your hard work in action.

13. Build and run your application to take a look at the results. When you select a station, you'll see a list of predictions that have been loaded for the station, as shown in Figure 7-20. Note that you may not see any predictions if no trains are currently in service.

Figure 7-20. *Viewing the predictions for the selected station*

You'll quickly notice a huge problem. We have no idea what the destination is for each train displayed in the predictions. Since the default table view cell only has a single label, we're not able to display as much information as we'd like. Creating a custom table view cell solves this problem.

14. Let's create an empty user interface XIB file for our new table view cell. In Xcode, select the Resources folder; then go to **File ➤ New**, and create an empty XIB file called *PredictionCell.xib*, as shown in Figure 7-21.

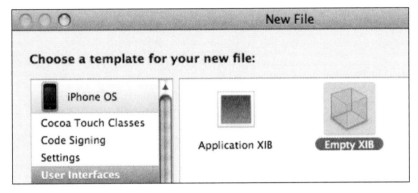

Figure 7-21. *Creating a new empty XIB for the custom table view cell*

15. You'll also need to create a class to go along with your new table view cell. Since we're customizing the prediction controller cell, select the Classes folder, go to **File ➤ New**, and create a subclass of UITableViewCell called PredictionCell, as demonstrated in Figure 7-22.

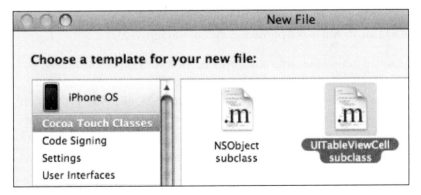

Figure 7-22. *Creating a new UITableViewCell subclass for the custom table view cell*

16. Next, we'll set up `PredictionCell` with a two outlets that we'll design in Interface Builder shortly. We'll use one label for the destination name and one label for the estimate of when the train will arrive. Again, you'll set up both instance variables as properties prefixed with the `IBOutlet` qualifier so that you can connect the labels in Interface Builder to the properties in the class. The code for our new class is shown in Listing 7-10.

Listing 7-10. *Creating the Header for the Custom Prediction Table Cell*

```
//
//  PredictionCell.h
//

#import <UIKit/UIKit.h>

@interface PredictionCell : UITableViewCell {
    UILabel *destinationLabel;
    UILabel *estimateLabel;
}

@property (nonatomic,retain) IBOutlet UILabel *destinationLabel;
@property (nonatomic,retain) IBOutlet UILabel *estimateLabel;

@end
```

17. Now, let's design our custom table view cell. Save any unsaved files in your project in Xcode, and then double-click *PredictionCell.xib* to open the file in Interface Builder. Drag an empty `Table View Cell` object from the library into your XIB. You should also associate your new empty table view cell with the class you created, `PredictionCell`, as shown in the properties dialog in Figure 7-23.

18. Now, we're ready to design the table cell. Double-click Prediction Cell, drag two labels to the content area of the table cell, and connect the outlets you created on the class to each of the labels, as demonstrated in Figure 7-24. Also note in the figure that the style of the top label has been changed using the **Font** menu to help distinguish the name of the destination from the prediction text. Feel free to experiment with the text styles to get the look that most appeals to you.

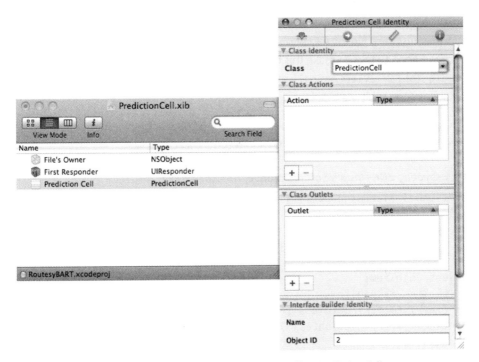

Figure 7-23. *Setting the class for the custom table view cell to PredictionCell*

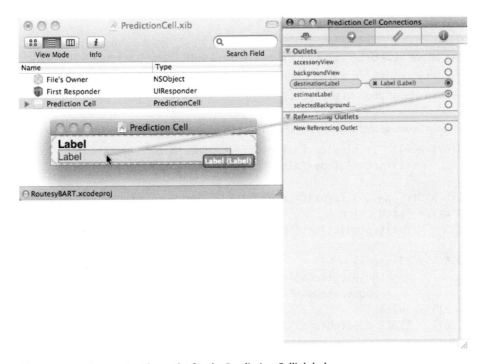

Figure 7-24. *Connecting the outlet for the Prediction Cell's labels*

19. Next, we need to configure `PredictionTableViewController` to use your new user interface file instead of the default table view cell. There are two steps we need to take to make this happen. First, add a new method to `PredictionTableViewController` called `createNewCell` to load a new instance of the table view cell.

As you can see in Listing 7-11, the code loads *PredictionCell.xib*, iterates through the items to find the first object of type `PredictionCell`, and returns that cell object.

Listing 7-11. *Creating a New Instance of PredictionCell from the XIB File*

```
- (PredictionCell*)createNewCell {
    PredictionCell *newCell = nil;
    NSArray *nibItems =
            [[NSBundle mainBundle] loadNibNamed:@"PredictionCell"
                                    owner:self options:nil];
    NSObject *nibItem;
    for (nibItem in nibItems) {
      if ([nibItem isKindOfClass:[PredictionCell class]]) {
        newCell = (PredictionCell*)nibItem;
        break;
      }
    }
    return newCell;
}
```

20. Now, all we have to do is change the `cellForRowAtIndexPath` method, which will now use your new cell and will set the text in the labels for both the destination and the estimate, as you can see in Listing 7-12.

Listing 7-12. *Customizing cellForRowAtIndexPath to Set the Cell Contents*

```
- (UITableViewCell *)tableView:(UITableView *)tableView
            cellForRowAtIndexPath:(NSIndexPath *)indexPath {

    static NSString *CellIdentifier = @"prediction";
    Prediction *prediction =
            [self.predictions objectAtIndex:indexPath.row];

    PredictionCell *cell =
                (PredictionCell*)[tableView
                  dequeueReusableCellWithIdentifier:CellIdentifier];
    if (cell == nil) {
        cell = [self createNewCell];
    }
```

```
    cell.destinationLabel.text = prediction.destination;
    cell.estimateLabel.text = prediction.estimate;

    return cell;
}
```

Now, when you build and run your application, you'll see a much better view of the predictions complete with destinations listed, like the one shown in Figure 7-25.

Figure 7-25. *Viewing predictions using the new custom Prediction Cell*

Adding Location-Based Information to Routesy

One of the major features that will set our project apart from other transit prediction solutions is the ability of the application to detect which station is closest to the user, since that station is the one the user is most likely to want to travel from. The iPhone SDK provides the

Core Location framework for determining the user's location using GPS or cell tower triangulation (see Figure 7-26). We'll get the user's location once when the application is launched and use that information to sort the list of stations.

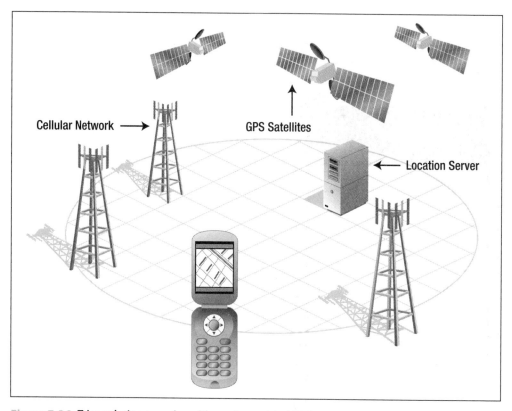

Figure 7-26. *Triangulating a user's position using assisted GPS*

1. First, we need to create a shared instance of CLLocationManager, the object that exposes the user's current location to our application. First, make sure you add an #import statement to *RoutesyBARTAppDelegate.m* to include the Core Location framework:

```
#import <CoreLocation/CoreLocation.h>
```

Then, add an instance of CLLocationManager to your application delegate, as shown in Listing 7-13, so that any part of the application can gain access to the location data. We'll also tell the location manager class that we'd like the accuracy to be within the nearest hundred meters, if at all possible.

TIP

> Requesting a high level of accuracy from Core Location has the price of increased battery drain on the device, so you should take that into account when building location-based applications. Your application should only request location updates when necessary and should stop updating once an acceptable location has been obtained.

Listing 7-13. *Adding a Shared CLLocationManager to RoutesyBARTAppDelegate*

```
+ (CLLocationManager*)sharedLocationManager {
  static CLLocationManager *_locationManager;

  @synchronized(self) {
    if (_locationManager == nil) {
      _locationManager = [[CLLocationManager alloc] init];
      _locationManager.desiredAccuracy = kCLLocationAccuracyHundredMeters;
    }
  }
  return _locationManager;
}
```

2. Next, you'll need to tell your instance of CLLocationManager that you'd like to start updating the user's location information. Put this code into the viewDidLoad method of RootViewController so that when the station list initially loads, we also begin attempting to locate the user.

3. CLLocationMethod also accepts a delegate object. In this case, we'll set the delegate property to self so that the RootViewController is notified when the user's location changes:

```
CLLocationManager *locationManager =
                 [RoutesyBARTAppDelegate sharedLocationManager];
locationManager.delegate = self;
[locationManager startUpdatingLocation];
```

4. CLLocationManager provides a didUpdateToLocation delegate method that is called whenever the user's location changes. Since we want to sort the list of stations by the closest one to the user, we'll have to implement this method in RootViewController so that the application can sort the list of stations once a location is obtained.

You'll also need to reload the table view after finishing the sorting, so that the table view's display is updated, as shown in the following code snippet. You'll notice that the following code stops updating the location with a call to stopUpdatingLocation on the location manager. That's because we've already obtained the user's location, and stopping the updates when they're no longer needed helps save battery life on the device.

```
- (void)locationManager:(CLLocationManager *)manager
     didUpdateToLocation:(CLLocation *)newLocation
                  fromLocation:(CLLocation *)oldLocation {
  [self sortStationsByDistanceFrom:newLocation];
  [self.tableView reloadData];
  [manager stopUpdatingLocation];
}
```

5. There's one more technicality that you'll need to deal with. Since our code now claims that RootViewController acts as a CLLocationManagerDelegate, we'll need to specify that in the protocol section of *RootViewController.h*. Change this declaration:

```
@interface RootViewController : UITableViewController
                               <BARTPredictionLoaderDelegate>
```

to this declaration

```
@interface
   RootViewController : UITableViewController
                        <BARTPredictionLoaderDelegate,
                        CLLocationManagerDelegate>
```

6. Now that the code is in place to determine the current location, we need to determine how far each station is from the user's position and sort the list of stations using that distance. We've already referenced a method that we haven't created yet, sortStationsByDistanceFrom. Let's implement this method now; it's shown in Listing 7-14.

Listing 7-14. *Sorting the Stations by Distance from a Location*

```
- (void)sortStationsByDistanceFrom:(CLLocation*)location {
  Station *station;
  CLLocation *stationLocation;
  for (station in self.stations) {
    stationLocation = [[CLLocation alloc] initWithLatitude:station.latitude
longitude:station.longitude];
    station.distance =
          [stationLocation getDistanceFrom:location] / 1609.344;
```

```
    [stationLocation release];
  }

  NSSortDescriptor *sort =
                  [[NSSortDescriptor alloc] initWithKey:@"distance"
                                              ascending:YES];
  [self.stations sortUsingDescriptors:[NSArray arrayWithObject:sort]];
  [sort release];
}
```

For each station in the list, we first initialize a new CLLocation object based on the station's latitude and longitude.

CLLocation has a method called getDistanceFrom that will return the number of meters of distance between two CLLocation objects. Since our application will be working with miles instead of meters, we can divide the distance by 1,609.344, which is the number of meters in a mile. Once we've calculated the distance, we set it into the station's distance property, which is the field that we'll use to sort the list.

That's where NSSortDescriptor comes in. Basically, this Cocoa class allows you to create a definition of how you'd like to sort your array. The NSSortDescriptor we create here will tell sortUsingDescriptors to sort the array using the distance property in an ascending order. Obviously, the station with the shortest distance is the one closest to the user, which means that the nearest station will appear at the top of the list.

NOTE

You may be wondering how the iPhone Simulator handles location detection. Apple has cleverly hard-coded the location of the simulator to 1 Infinite Loop in Cupertino, California—the location of Apple's corporate headquarters.

If you build and run your application, you'll see that a second or two after the application launches, your list of stations is sorted, with Fremont at the top, since this is the closest location to your simulator.

Putting the Finishing Touches on Routesy BART

So far, you've built a reasonably useful application that serves its basic purpose: telling a user when the next train will arrive at the nearest station. However, there are a few easily implementable details that will add a bit of extra usefulness to Routesy. The default cell style on the station table isn't particularly helpful. It would be useful to show the user how far away the nearest station is.

Let's finish by creating a custom table cell for the station table view. If these steps seem familiar, it's because they are the same steps we took to create the custom cell for the predictions table.

1. First, let's create a class for the custom table view cell. The code for this class is shown in Listing 7-15. This cell will display a station, so call it StationCell, and place it in the View folder. Remember that Xcode provides you with a helpful template for creating subclasses of UITableViewCell. The code for this class is very simple. Create and synthesize two properties, stationNameLabel and distanceLabel, one for each of the values that the cell will display, and make sure to place an IBOutlet declaration before the type in each property so that you can connect the labels in Interface Builder to your class.

Listing 7-15. *Creating a Class for the Custom Station Table Cell*

```
//
//  StationCell.h
//

#import <UIKit/UIKit.h>

@interface StationCell : UITableViewCell {

    UILabel *stationNameLabel;
    UILabel *distanceLabel;

}

@property (nonatomic,retain) IBOutlet UILabel *stationNameLabel;
@property (nonatomic,retain) IBOutlet UILabel *distanceLabel;

@end

//
//  StationCell.m
//

#import "StationCell.h"

@implementation StationCell
@synthesize stationNameLabel, distanceLabel;

- (void)dealloc {
    [stationNameLabel release];
```

```
    [distanceLabel release];
    [super dealloc];
}

@end
```

2. Next, create an Interface Builder XIB file for your new cell called *StationCell.xib*, and place it in your resources folder. Again, you'll use an empty XIB file and place one cell from the library in the file. Set the cell's identifier to "station" and the class to the `StationCell` class you created previously, as shown in Figure 7-27, the same way you did for `PredictionCell`. Make sure your cell attributes are identical to the ones in the image shown in Figure 7-27.

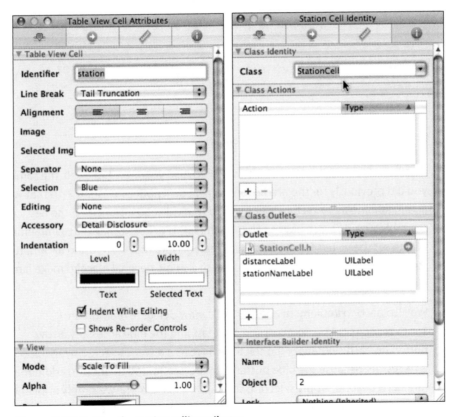

Figure 7-27. *Setting the station cell's attributes*

3. Now, drag two labels into your table cell, as shown in Figure 7-28: one for the station name and one for the distance. Connect the two outlets to your labels so that you'll be able to access them from your code.

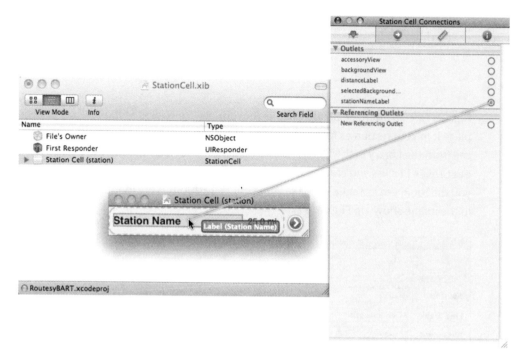

Figure 7-28. *Connecting the station cell's label outlets*

4. Finally, we'll set up `RootViewController` to load your new table cell, the same way you did previously for the predictions table. You'll notice in Listing 7-16 that we're again defining a `createNewCell` method that will load instances of the new cell into the table view. We also have to change the `cellForRowAtIndexPath` implementation to set the values for the fields we created. Note that we're using a formatted string to round the digits of the distance to a single decimal place to make it more display-friendly.

We also need to implement another table view delegate method on `RootViewController` to make sure that tapping the blue accessory arrow button on each cell causes the cell to behave as if it were selected by the user. This is necessary because the blue buttons can be configured to perform additional functions beyond simply selecting the cell. In our example, we simply implement `accessoryButtonTappedForRowWithIndexPath` to select the cell programmatically.

Listing 7-16. *Setting Up the Code for the Custom Station Cell*

```
- (UITableViewCell *)tableView:(UITableView *)tableView
        cellForRowAtIndexPath:(NSIndexPath *)indexPath {
  static NSString *CellIdentifier = @"station";
  Station *station = [self.stations objectAtIndex:indexPath.row];
```

```
StationCell *cell =
    (StationCell*)[tableView
                    dequeueReusableCellWithIdentifier:CellIdentifier];
if (cell == nil) {
      cell = [self createNewCell];
}

cell.stationNameLabel.text = station.name;
if (station.distance) {
  cell.distanceLabel.text = [NSString stringWithFormat:@"%0.1f mi",
                                        station.distance];
} else {
  cell.distanceLabel.text = @"";
}

return cell;
}

- (StationCell*)createNewCell {
  StationCell *newCell = nil;
  NSArray *nibItems =
      [[NSBundle mainBundle]
                  loadNibNamed:@"StationCell"
                        owner:self options:nil];
  NSObject *nibItem;
  for (nibItem in nibItems) {
    if ([nibItem isKindOfClass:[StationCell class]]) {
      newCell = (StationCell*)nibItem;
      break;
    }
  }
  return newCell;
}
- (void)tableView:(UITableView *)tableView
      accessoryButtonTappedForRowWithIndexPath:
          (NSIndexPath *)indexPath {
    [self tableView:tableView didSelectRowAtIndexPath:indexPath];
}
```

When you build and compile your project, your application will open to a much nicer view for the user, like the one shown in Figure 7-29, complete with the distance listed for each station.

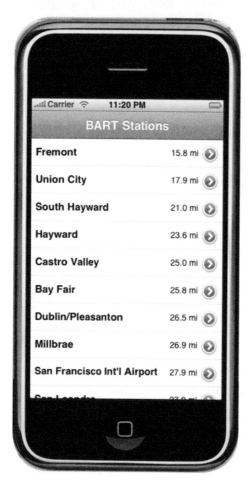

Figure 7-29. *Viewing the station list with location information*

Summary

In this chapter, you've seen how to take an external data source and utilize it in a location-based application to create a handy utility for iPhone users. Routesy is only one example of how the vast data resources on the Internet can be put into the context of location and always-on networking to provide information to people in more rich and interactive ways than ever before.

Your sample Routesy transit prediction application brought together several of the iPhone's core technologies. The topics covered included:

- Building a hierarchical set of views using `UINavigationController` to allow the user to move back and forth between screens

- Populating a `UITableView` using dynamically loaded data

- Building custom `UITableViewCell` views using Interface Builder

- Using `NSURLConnection` to asynchronously load data from the Internet

- Parsing data loaded from an XML feed

- Using Core Location to locate the user and sort nearby places

If you find yourself building applications with location-based capabilities, hopefully you will find pieces of this application example useful. Many of the features we implemented are common in iPhone applications: from sorting data by location, to loading content from online data sources. With each application you build, these tasks will become more familiar, and as a result, easier. Good luck!

Index

You Need the Companion eBook

Your purchase of this book entitles you to buy the companion PDF-version eBook for only $10. Take the weightless companion with you anywhere.

We believe this Apress title will prove so indispensable that you'll want to carry it with you everywhere, which is why we are offering the companion eBook (in PDF format) for $10 to customers who purchase this book now. Convenient and fully searchable, the PDF version of any content-rich, page-heavy Apress book makes a valuable addition to your programming library. You can easily find and copy code—or perform examples by quickly toggling between instructions and the application. Even simultaneously tackling a donut, diet soda, and complex code becomes simplified with hands-free eBooks!

Once you purchase your book, getting the $10 companion eBook is simple:

1. Visit **www.apress.com/promo/tendollars/**.

2. Complete a basic registration form to receive a randomly generated question about this title.

3. Answer the question correctly in 60 seconds, and you will receive a promotional code to redeem for the $10.00 eBook.

THE EXPERT'S VOICE™

2855 TELEGRAPH AVENUE | SUITE 600 | BERKELEY, CA 94705

Offer valid through 01/2010.